Social Licensing and Mining in South Africa

This book highlights the role of community trusts in social licensing through the lens of mining and mining disputes in South Africa.

Employing elements of trust, acceptance, and elite interaction as a framework, this book critically investigates the underlying dynamics of community development trusts and also the response of host communities to the inherent dilemma of the social licence to operate (SLO) concept, namely social legitimation versus corporate profits. Looking at formal versus informal regulatory requirements, popular mobilisation, and the interaction between the local population and mining companies, this book constitutes a thorough look at the issues surrounding mining in South Africa and its effect on society.

This book will be of interest to students and scholars of African studies, business in Africa, corporate responsibility, and development studies.

Sethulego Matebesi is a Senior Lecturer in the Department of Sociology at the University of the Free State (UFS), South Africa.

Routledge Contemporary Africa Series

Life-Writing from the Margins in Zimbabwe
Versions and Subversions of Crisis
Oliver Nyambi

Complex Adaptive Systems, Resilience and Security in Cameroon
Manu Lekunze

The International Criminal Court and the Lord's Resistance Army
Enduring Dilemmas of Transitional Justice
Joseph Otieno Wasonga

African Intellectuals in the Post-colonial World
Fetson A Kalua

National Healing, Integration and Reconciliation in Zimbabwe
Edited by Ezra Chitando

Africa's Elite Football
Structure, Politics, and Everyday Challenges
Edited by Chuka Onwumechili

Social Licensing and Mining in South Africa
Sethulego Matebesi

The Everyday Life of the Poor in Cameroon
The Role of Social Networks in Meeting Needs
Nathanael Ojong

For more information about this series, please visit: www.routledge.com/Routledge-Contemporary-Africa/book-series/RCAFR

Social Licensing and Mining in South Africa

Sethulego Matebesi

LONDON AND NEW YORK

First published 2020
by Routledge
4 Park Square, Milton Park, Abingdon, Oxon OX14 4RN
605 Third Avenue, New York, NY 10017

First issued in paperback 2023

Routledge is an imprint of the Taylor & Francis Group, an informa business

© 2020 Sethulego Matebesi

The right of Sethulego Matebesi to be identified as author of this work has been asserted by him in accordance with sections 77 and 78 of the Copyright, Designs and Patents Act 1988.

All rights reserved. No part of this book may be reprinted or reproduced or utilised in any form or by any electronic, mechanical, or other means, now known or hereafter invented, including photocopying and recording, or in any information storage or retrieval system, without permission in writing from the publishers.

Trademark notice: Product or corporate names may be trademarks or registered trademarks, and are used only for identification and explanation without intent to infringe.

British Library Cataloguing-in-Publication Data
A catalogue record for this book is available from the British Library

Library of Congress Cataloging-in-Publication Data
A catalog record has been requested for this book

ISBN: 978-1-03-257058-7 (pbk)
ISBN: 978-1-138-34893-6 (hbk)
ISBN: 978-0-429-43107-4 (ebk)

DOI: 10.4324/9780429431074

Typeset in Baskerville
by codeMantra

Publisher's Note
The publisher has gone to great lengths to ensure the quality of this reprint but points out that some imperfections in the original copies may be apparent.

To Lesego, Puleng and Tshiamo

To Laeego, Paieng and Tshiamo

Contents

List of figures	ix
List of tables	xi
List of photographs	xiii
Acknowledgements	xv
List of abbreviations	xix

1 Introduction: context of community–mine relations in South Africa — 1

2 Path dependency and social licensing — 25

3 Mining regulatory frameworks and civil society mobilisation — 52

4 Royal Bafokeng Nation: a model of community-based natural-resource management — 83

5 Social mobilisation against community development trusts in South Africa — 110

6 Conclusion: social licensing and prospects for community development — 139

Index — 151

Figures

1.1 Map of South Africa and locations of case studies 15
4.1 Map of Rustenburg and surrounding areas 85

Tables

3.1	Overview of pre-1994 mining legislation in South Africa	54
3.2	Overview of post-apartheid mining legislation in South Africa	58
4.1	Selected timeline of historical events related to RBN	87
4.2	Procedural fairness and trust in RBNDT (percentage in parentheses)	99
4.3	RNBDT's contribution to the quality of life of Phokeng residents (percentage in parentheses)	100
4.4	Future behavioural intention regarding selected variables (percentage in parentheses)	102
5.1	Procedural fairness and trust in CDT (percentage in parentheses)	130
5.2	CDT's contribution to the quality of life of residents (percentage in parentheses)	131
5.3	Future behavioural intention regarding selected variables (percentage in parentheses)	131

Photographs

Photo 1   Roads blocked with rocks in Atok   xxi
Photo 2   Atok village, Limpopo   xxi
Photo 3   Jagersfontein protestors in front of burned town hall   xxii
Photo 4   Jagersfontein residents marching to mine and CDT   xxii

Acknowledgements

The present book is based on a three-year (2016–2018) research project of community development trusts and conflict in mining-affected communities in four provinces of South Africa. However, its genesis can be traced to my former role as Research Associate at the Centre for Development Support (CDS) of the University of the Free State (UFS). I was part of a research team of the CDS that was commissioned by the Centre for Enterprise Development – a Johannesburg-based independent public policy research and advocacy non-profit organisation established in 1995 – to conduct four case studies at the municipal level in 2006 on what is popularly known as "service delivery protests". The project captured my attention and piqued my curiosity in such a way that I began a three-year study of the widespread grassroots forms of mobilisation against municipalities between 2013 and 2015, which was funded by the Thuthuka Programme of the National Research Foundation. This research project led to my extended research visit to Uppsala, Sweden in 2014, which ultimately culminated in the publication of *Civil Strife against Local Governance: Dynamics of Community Protests in Contemporary South Africa* in 2017. Two major themes for future research emerged from this project: how core protest leaders benefit from incentives provided by the political or social elites (I call this *transactional activism*), and the role of community development trusts in fomenting community protests.

My interest in mining–community relations was further shaped by my mentor, Professor Lochner Marais of the Centre for Development Enterprise, who, when I shared my future research workstreams with him, told me, "Zachy, believe me, you will have scope for work for the next 20 years within the extractive field." It is thus no surprise that I begin to express my profound gratitude to Prof. Marais for being mostly responsible for shaping my scholarly journey. I benefited immensely from our monthly interaction over the past two years as a mentor for the Andrew Mellon Foundation Inclusive Professoriate Programme at UFS. Furthermore, I am immensely impressed by Prof Marais' relentless efforts to promote research excellence with policy relevance in the service of society.

I would also like to thank Routledge Publishers for selecting my book for publication and, as a result, providing a voice to the many voiceless mining-affected communities in South Africa and across the world. Thanks to Leanne Hinves and Henry Strang – the incredibly knowledgeable editorial team of the

Routledge African Studies list – who were my main points of contact during the writing stage and for preparing the submitted manuscript for the production team. My sincere appreciation goes to the anonymous peer reviewers who reviewed the book proposal as well as the completed manuscript.

Social Licensing and Mining in South Africa would not have been possible without generous funding from the Thuthuka Programme of the National Research Foundation of South Africa, as well as the Andrew Mellon Foundation Inclusive Professoriate Programme hosted by the Faculty of Humanities of UFS. A special thanks to Dean of the Faculty of the Humanities, Prof. Heidi Hudson, and former deans, Prof. Lucius Botes and Prof. Milagros Rivera, for creating a rich, stimulating, and engaging environment for the Mellon Fellows. I would also like to thank Marizanne Cloete for the excellent manner in which she dealt with the administration of the Mellon programme. I am also grateful to the entire Mellon Fellows for the vibrant and intense academic deliberations, which contributed to my intellectual growth. I also extend my thanks to Billy Kalima (doctoral candidate) and Siyanda Magayana (master's candidate) whom I had the privilege of supervising as part of the Andrew Mellon Foundation scholarship.

The writing of this book has been a fantastic journey that took me to Australia and Canada, two of the world's top mining countries. At Queensland University, Prof. Deana Kemp, Prof. John Owen, and Dr Kathryn Stuurman of the Centre for Social Responsibility in Mining (CSRM) deserve my gratitude for being my hosts and reinvigorating my enthusiasm each time I felt disheartened. I wish to single out the contribution made by John for reading the entire completed manuscript. John's advice has been enormously helpful, and has broadened and strengthened the arguments in this book. A special word of thanks also goes to Dr Jane Thomson of the Emmanuel College in Queensland, for providing an academically conducive environment. At the University of Alberta, I want to thank the Department of Earth and Atmospheric Sciences for having accepted my request to host me as a visiting researcher. I acknowledge, in particular, the support of Prof. Kristof Van Assche and his wife, Monica Gruezmacher, who welcomed me and ensured that my stay in Edmonton is both enjoyable and academically enriching.

I also feel incredibly privileged to have had Bruce Conradie to manage the language editing of the draft manuscript. I could not have asked for a better person with an eye for detail and consistency in style. The insights and experience of the Mining and Society Study Group – a multidisciplinary group across faculties and universities – were of importance in sharpening the ideas in this book.

I am also grateful to Dolly Mashele, Karabo Gaearwe, Masabata Khaile, Moleleki Motaung, and Pontso Moletsane who sacrificed their time and had to endure severe conditions during data collection. I am especially grateful to the many participants for their candour in expressing their lived experiences featured in Chapters 4 and 5. I benefited tremendously from my role as a commentator on current affairs in South Africa through various media platforms, in particular, SABC TV, Lesesedi FM, and Motsweding FM.

Finally, my sincerest gratitude goes to my family for their support. My wife, Lesego, has been a source of constant and unconditional love. I am grateful that she allowed me to venture on numerous research trips and to manage the fort in my absence. To my daughter, Puleng, and son, Tshiamo, thank you for your enthusiasm despite the enormous strain my absence placed on your daily routines. It is almost impossible to describe the heart-wrenching moment each time I had to say my final goodbye before departing on international sojourns. The memories of those moments, I reckon, will last a lifetime.

Sethulego
Edmonton, Canada
July 2019

Abbreviations

ACC	Amadiba Crisis Committee
AMI	Alternative Indaba of Mining
AMV	Africa Mining Vision
ANC	African National Congress
AU	African Union
BBBEE	Broad-Based Black Economic Empowerment
BBNCEC	Baroka Ba Nkwana Community Engagement Committee
BEE	Black Economic Empowerment
BLBA	Bafokeng Land Buyers' Association
CDAs	Community Development Agreements
CSOs	Civil Society Organisations
CSR	Corporate Social Responsibility
DMR	Department of Mineral Resources
DRC	Democratic Republic of Congo
EGT	Evolutionary Governance Theory
EIA	Environmental Impact Assessment
EITI	Extractive Industries Transparency Initiative
FPIC	Free, Prior, and Consent
FTLR	Fast Track Land Reform
IBMR	Itireleng Bakgatla Mineral Resources
ICMM	International Council on Metals and Minerals
ICT	Itumeleng Community Trust
IDP	Integrated Development Plan
ILO	International Labour Organisation
ILUAs	Indigenous Land Use Agreements
IPILRA	Interim Protection of Informal Land Rights Act
JMMF	Joint Multi-Stakeholder Mining Forum
LLAs	Local-Level Agreements
MACUA	Mining Communities United in Action
MPRDA	Mineral and Petroleum Resources Development Act
NAFTA	North American Free Trade Agreement
OECD	Organisation for Economic Cooperation and Development
PNG	Papua New Guinea

PPM	Pilanesberg Platinum Mines
RBA	Royal Bafokeng Holdings
RBH	Royal Bafokeng Holdings
RBN	Royal Bafokeng Nation
RBNDT	Royal Bafokeng National Development Trust
SIOC	Sishen Iron Ore Company
SLO	Social Licence to Operate
TEM	Transworld Energy Minerals
UN	United Nations
WAMAU	Women Affected by Mining United in Action

Photo 1 Roads blocked with rocks in Atok.

Photo 2 Atok village, Limpopo.

Photo 3 Jagersfontein protestors in front of burned town hall.

Photo 4 Jagersfontein residents marching to mine and CDT.

1 Introduction

Context of community–mine relations in South Africa

Introduction

One of the most critical questions facing resource-rich countries globally today is how to optimise natural-resource revenues for the long-term sustainable development of mining communities. Consequently, it is becoming increasingly necessary for mining companies across the world to adhere to the requirements of a social licence to operate (SLO). This book examines community development trusts ('CDTs', or community trusts) established voluntarily by mining companies in South Africa to foster social licensing and, as a consequence, minimise conflict. A lack of community acceptance – a fundamental dimension of social licensing – is a significant risk for the interruption, postponement, and shutdown of mining projects.

Over the years, mining companies across the world have sought ways to minimise conflict in their interactions with local communities. The factors that primarily provoke opposition to or acceptance of mining projects are most often related to the mining company's ability to ensure local responsibility, sustainability, and accountability. Today, local communities, such as Australia, Canada, Kenya, Niger, the Philippines, and Tanzania, demonstrate a growing unease over their exclusion from the decision-making process of natural-resource governance that affects them.

The conflict between communities and mining companies in South Africa is rooted in the history of apartheid, which allowed the government and the mining industry to act unilaterally. In explaining attempts by the post-apartheid government to address the conflict, some scholars identify two policy directives closely associated with social licensing: social and labour plans and Mining Charters. First, the government introduced social and labour plans to address the need for collaborative planning at the local level in 2002 (Marais, 2013). These plans set out how the company intends to share some of the benefits that flow from mining, including, for example, developing employee skills, upgrading local schools and roads, and providing housing, water, and sanitation to the mining community (Centre for Applied Legal Studies, 2017).

Second, the government has been issuing documents and policy guidelines – known as Mining Charters (Suttner & Cronin, 1986; Ramatji, 2013; Rametsi,

2016; DMR, 2017) – to promote the idea of local ownership in mining operations, thereby engendering a form of social licensing. For example, on 27 September 2018, the Minister of Mineral Resources published the 'Broad-Based Socio-Economic Empowerment Charter for the South African Mining and Minerals Industry, 2018' (the Mining Charter, 2018), which envisages that workers and the mining community will have at least an 8% share in the local operations of mining companies (DMR, 2018).

Interestingly, outside this policy framework, some mining companies have, on their own initiative, created community trusts with 2–5% local shareholding, all in line with the idea of creating an SLO (Matebesi & Marais, 2018). Ironically, as I will explain throughout this book, these community trusts have been a significant source of conflict in mining communities in South Africa. The conflict, I argue, is not only an iteration of the ongoing and widespread protests across South Africa, but also an indication of various unresolved historical concerns over land rights and mining. For example, since 2001, the Driekop village in the Limpopo Province has regularly clashed with Modikwa mine over the right to use farmlands near the mine. This village lies in the district with the most significant platinum reserves in the world; yet it remains one of the most poverty-stricken areas in South Africa (ActionAid, 2008). With the headline "Community blames 'arrogant' mine for unrest," the *City Press* newspaper provided a chilling account of the killing of six workers of Modikwa mine when a bus transporting them to their night shift was petrol-bombed on the evening of 2 April 2018. Community leaders accused the management of the mine of ignoring the storm that had been brewing for a while, and of refusing to meet the community members (Malope, 2018).

My purpose in this book is to highlight how community trusts play a central role in social licensing by seeking answers to several questions: What are the governance structures and community engagement processes of the community trusts? Do they create a social acceptance of mining? Do they make any meaningful difference in how mining communities view the social benefits of mining and engagement? What can historical and other sources of insight offer to the future of community trusts?

In the light of these questions, this introductory chapter provides a general overview of social licensing, community trusts, and the theory underpinning the study. It should be noted that the context of South Africa discussed in this chapter has roots elsewhere in the world, particularly in countries where historical government policies impact on mining–community relations. This chapter enables readers to cast the significance of social licensing in historical terms, and provides insight into the paradoxes of resource endowment and human development in developing countries.

The next section discusses the institutional context of mining by focusing on participatory governance. This is followed by a description of mining and civil society struggles in advocating for mining communities to claim and exercise their rights to the benefits of mining. Next, I focus on community trusts in the form of local-level agreements (LLAs). The subsequent section presents a brief discussion of the literature on social licensing and the path dependency within evolutionary governance theory (EGT), which underlies the cases in the study. The last two sections provide the rationale for the book, followed by an outline of its structure.

Community–mine relations in South Africa – a rocky road

Internationally, the social responsibility agenda of mining has received considerable attention (IIED, 2002; World Bank, 2012; ICMM, 2013; Martineza & Franks, 2014; Vivoda et al., 2017). The central message of these divergent international voices is that local communities are fundamental to the long-term sustainability of mining. However, in 2011, a report by the United Nations Economic Commission for Africa (UNECA, 2011) bemoaned that, more than 40 years after independence, mining in Africa still exhibited colonial structural features. The UNECA report argued that the historical structural deficiencies of mining are characterised by enclaves of mining activity with fragile links to local economies.

Similarly, the history of mining in South Africa is characterised by a biased narrative that focuses on the performance of white-owned mining companies (Davenport, 2017). Other features of mining history include, for example, the exploitation of black mineworkers, the systemic exclusion of local communities (Rutledge, 2017), and the establishment of mining towns by some mining companies (Marais et al., 2016), as well as the industry's framing of the community as agentless subjects (Rutledge, 2017). Consequently, mining in South Africa is characterised by long periods of instability and conflict that echo class and race relations in the country (Busacca, 2013). In a terse conclusion about the political and social context of mining, Harvey (2017: 3) notes: "Racialised inequality, injected into the fibre of South Africa's economic trajectory through colonialism and apartheid, continues to undermine the prospects for black South Africans to escape poverty and build wealth." He further asserted that "combined with contemporary governance ineptitude, characterised by the acquisition of rents through state-owned entities and the distribution of those rents to politically-connected insiders, the prospects for most of the country's citizens are bleak" (Harvey, 2017: 3).

The recently heightened activism of civil society organisations against the mining industry and the government has brought participatory governance back into the spotlight. Broadly, attempts by post-apartheid South Africa to promote a more egalitarian society are underscored by several laws and policies aimed at building democratic and transparent public participation processes at all levels of government. The trend in the government's approach to participatory governance at the local government level has been threefold. First is the redefinition of the term "municipality" to include the notion of "local community" (see Thompson et al., 2018, for a more nuanced view on the conceptualisation of "community in the context of local governance," alongside others in the legal definition of a municipality). Second is the establishment of a set of requirements for public involvement in various decision-making processes and as a precondition for the receipt of central government grants (Thompson et al., 2018). For example, the preparation and adoption of integrated development plans (IDPs) require the broad-based participation of the local population. These IDPs are set up for five years, and participation is based on stakeholder meetings to discuss priorities (Terwindt & Schliemann, 2016).

The third innovation is the establishment of ward committees in politically demarcated municipal wards. The ward committees consist of the politically elected councillor, and approximately ten people from diverse interest groups in the ward (Piper & Nadvi, 2010). Despite the extensive legal framework that enabled the development of the local government participatory governance paradigm, the culture of genuine public involvement and engagement remains elusive (Murambo, 2008; Matebesi, 2017; Modise, 2017).

Globally, scholars have pointed to the ways in which participatory governance is used in sectors such as energy (Lariña et al., 2011; Xavier et al., 2017), construction of infrastructure (Groves et al., 2013), and mining (Terwindt & Schliemann, 2016; Conde, 2017). However, in countries such as India, Mexico, the Philippines, and South Africa, laws and administrative decisions allow for and foster natural-resource extraction without ensuring adequate participation rights for local populations (Terwindt & Schliemann, 2016). Leonard (2017: 331) argued that, in South Africa, "corporate influence over government has influenced state decision-making and how citizens have been drawn into participatory processes to inform decisions."

For example, despite being legally binding, social and labour plans are to be developed in consultation with a mining community, including other stakeholders such as mine workers and the local municipality. The plans are developed in a five-year cycle to coincide with the IDPs of the local municipality. In practice, the social and labour plan guidelines are violated by some mining companies; or worse, those companies may even outsource the formulation of social and labour plan to consultants (Mosaval et al., 2017). However, how is this possible? Seemingly, the engagement of the African National Congress (ANC) in macroeconomic policies and the need to create jobs could, in part, be the reason why the Department of Mineral Resources, in some instances, approves mining applications without making informed decisions (Leonard, 2017).

A further layer of complexity in the area of participatory governance for mining communities concerns the attitude of the Department of Mineral Resources. Following the democratic political dispensation in South Africa in 1994, several regulatory reforms were implemented. The earliest indication of the government's intention to acknowledge local ownership of mining came with the Mineral and Petroleum Resources Development Act (MPRDA) No. 28 of 2002 (Department of Mineral Resources, 2002). The MPRDA is one of the primary laws that regulates mining in South Africa (Malesa & Jackson, 2017). However, it failed to be prescriptive with respect to local ownership. The new regulatory regime for the industry was further entrenched with the launch of the Mineral Policy in October 1998, which proposed that the state, as the legal owner of the resources, is eligible for mining royalties (Cawood & Minnitt, 2001).

Internationally, in countries such as Australia, Argentina, Ghana, Guinea, Indonesia, and Nigeria, a percentage of mining royalties are ring-fenced for the benefit of communities impacted by mining (Chabana, 2016). Since then, the notion of local ownership underwent considerable changes as it coincided with four mining charters (memoranda of understanding between the government and the mining industry) that were promulgated between 2004 and 2018. The 2018 Draft Mining Charter, gazetted in June 2018, envisaged that workers and the mining communities would have at least an 8% share in mining companies' local operations (DMR, 2017).

Generally, although institutional change is apparent in how mine–community relationships are managed, many of the historical concerns remain (Matebesi & Marais, 2018). This point is emphasised by Clapham (2005: 8), who argues that "much government policy is following and reinforcing established trends rather than seeking new directions." In this respect, one can argue that the democratisation process in South Africa has not yielded positive outcomes for mining communities. It is also apparent that policies alone have not been sufficient to address historical government failures. The sporadic attempts at mobilisation by grassroots organisations that are against being trapped in a system that continues the legacy of apartheid and dispossession (ActionAid, 2016), have achieved limited success until now. However, national civil society organisations, including, for example, the Mining Affected Communities United in Action South Africa (MACUA-SA), Women Affected by Mining United in Action (WAMUA), and the Mining and Environmental Justice Network of South Africa (MEJCON-SA), have managed to reframe traditional mining issues and shaped public discourse on the rights of mining communities.

2018 witnessed the most dramatic demonstration of the prominence of civil society, when MACUA-SA and many other national organisations won several court cases against the government, particularly over the issue of community engagement. For example, in February 2018, Pretoria High Court ordered the South African government and the Minerals Council South Africa (formerly Chamber of Mines) to recognise that all mining communities are core stakeholders who must be consulted in the formulation of any new Mining Charter (Bruce, 2018). Another milestone for civil society's struggle for recognition has been the Alternative Mining Indabas, a dialogue for civil society organisations that runs parallel to the Mining Indaba. The Ninth Alternative Indaba was held in February 2017 and attended by 350 organisations. The civil society organisations came to the critical conclusion that the trajectory of the country's mining industry was unsustainable and a radical shift is needed to maximise the benefits of resource extraction for all stakeholders, including mining communities (Adeleke, 2017).

Efforts by local communities to defend and reclaim their rights to livelihoods, food security, and environmental rights have been met with harassment, intimidation, or violence (Human Rights Watch, 2019). The ongoing systemic and institutional nature of the violations experienced by mining communities has the potential to erode the gains of civil society in asserting the right of communities to be involved in decisions about mining projects. For example, at the local government level, the deficit in social accountability mechanisms and the failure of municipalities to disclose information on what mining companies contribute to IDPs (Dlamini & Mbangula, 2017) pose devastating effects on community agency. These flawed regulatory and institutional arrangements that characterise the mining sector in South Africa, as well as in countries such as Armenia, Ghana, Guatemala, and Macedonia, have contributed to corruption and poor governance, as one mineworker confirms:

> Look chief, I was born and bred here. The first time I saw a top-of-the-range German machine [Mercedes-Benz] was with mine owners from Jozi [Johannesburg]. Naturally, I wanted to be like them. I went through the system and qualified; only to find that you need connections more than papers, so I connected myself, he said.
>
> (Malope, 2018: 5)

Another factor that may drastically erode the capacity of *civil society* is the continued business-centric views of some scholars. For example, research reports such as "Transformation Trumps Sustainability" (Jeffrey, 2018) are engaging, but also lend credence to Davenport's (2017) warning: "If there is one thing South African mining historians, myself included, have been guilty of, it is the tendency to focus on, and perpetuate, the 'colonial-esque' history of the industry." Only time will tell whether the recent surge in civil society activism in the mining sector in South Africa will yield the same results as in countries such as Mexico (Godoy, 2013).

Since the advent of democracy, state–civil society relations are crucial for understanding community, and mining communities in South Africa should also be seen in this context. In explaining the extensive process of state reform, aimed at building a more egalitarian social and political order, some scholars, for example Tapscott (2017), emphasised that progress has been made in reforming the state infrastructure; however, he continued, "the pursuit of an inconsistent economic growth path has meant that inequality and poverty remain serious challenges to the social order" (Tapscott, 2017: 69). According to Tapscott (2017: 76), "state-civil society relations remain less clearly defined," and the legacy of apartheid has left a racially intolerant and unequal society, which is preventing the country from establishing a social compact.

Globally, scholars such as Bunker (1989), Boyd et al. (2001), and Kirsch (2014) have contributed to a new wealth of knowledge on the geographies and political ecologies of resource extraction. This body of research chronicles local and regional transformations that brought about resource boom and bust cycles.

Community trusts as forms of LLAs

This section discusses mining–community relations in the context of community trusts. However, what exactly do I mean by a community trust? To avoid any ambiguity of the concept, I use the term to illuminate the particular problem of community trusts created voluntarily by mining companies in South Africa to manage royalty payments. Community trusts are community-based and work for the well-being and sustainable development of their area through economic, environmental, social, and cultural initiatives. Community trusts serve as the implementing structure of what are internationally known as LLAs or community development agreements (CDAs) (Brereton et al., 2011; O'Faircheallaigh, 2012; Keenan et al., 2016; Loutit et al., 2016). I use the terms "LLAs" and "CDAs" interchangeably. The term "mining community" is used as a proxy to denote the mining host community (in the immediate proximity of the mining project) and mining communities (communities outside the immediate proximity of the mine), as well as vested stakeholders.

Globally, there is an expanding body of research on mining company–community agreements, which are increasingly seen as a critical component of sustainable development in the context of mining (O'Faircheallaigh, 2004, 2010, 2011, 2012; Brereton et al., 2011; Keenan et al., 2016; Loutit et al., 2016). While early agreements in Australia and Canada focused mainly on compensation for land access and acquisition (Keenan et al., 2016), the transition has been moving to a focus on the absence of women (Brereton, 2011; O'Faircheallaigh, 2011), followed by a more recent focus on the legal and institutional frameworks (Nwapi, 2017). The drivers of CDAs over the past few decades have been globalisation, demands for accountability from mining stakeholders (Sarkar, 2010), and the emphasis on benefit sharing and the promotion of equity (O'Faircheallaigh, 2010). Voluntary programmes are the primary features of CDAs in the extractive industry, which are centred on meaningful community involvement (Loutit et al., 2016).

In practice, though, CDAs vary widely in application and impact. For example, Sarkar et al. (2010) pointed out that benefit sharing have been included in mining regulations of countries such as Chile, Papua New Guinea (PNG), and South Africa. In Africa, Ghana, Tanzania, the Democratic Republic of Congo (DRC), and Namibia are increasingly seeking to entrench community development initiatives within their policy frameworks (Sarkar et al., 2010). Interestingly, in Malawi, CDAs have evolved as both voluntary and obligatory tools to improve the relations between companies and communities. Etter-Phoya (2017) notes that new mining legislation in Malawi will require mining companies to spend at least 0.45% (far less than the 8% proposed by the latest mining charter in South Africa) of their annual gross revenue on community development activities.

In South Africa, mining companies have sought increasingly to gain and maintain community approval of their operations by initiating informal community-based implementation units called community development trusts (CDTs). Increasingly, these development trusts are becoming one of the critical instruments of corporate legitimation and social acceptability in the country. As the De Beers Group (2009: 67) stated in its Report to Society, "effective engagement is essential for constructive, frank and stable relationships with local communities and other interested parties." The report continued, "Without such relationships, our business risks operational disruption as well as reputational damage. These relationships also inform our efforts to generate lasting socio-economic benefits at a local level and to maintain our social licence to operate" (De Beers Group, 2009: 67).

The literature on social acceptance of mining is dominated by studies on the formal, government-stipulated community social responsibility initiatives of mining companies as regulated by social and labour plans in South Africa.

Six principal observations arise from the analysis of CDTs in South Africa. First, the trusts are structures independent of mining operations through which community equitable shares – usually between 2% and 5% of investment related to mining corporate social responsibility (CSR) – are directed. Second, though the trusts are conceived to be representative and participatory, unilateral decision-making seems to be a key feature of both their communicative and organisational approaches. Third, the local development initiatives of the trusts must be aligned with the local municipality's IDPs. The main challenge confronting the CDTs in terms of the alignment process is capacity constraints at the local government level. These constraints include the tension of dealing with divergent interests and demands from communities, elite (Sidley, 2015), as well as having to meet regulatory requirements.

Fourth, as Harvey (2017: 5) explained, an "unfortunate geographic reality is that mining-dependent local municipalities are mainly concentrated in the country's peripheral and poor provinces." As a result, most communal land tenure is typically held in tribal trusts. Such community trusts generally tend to advance the goals of traditional leaders rather than those of community development, thereby locking communities in pathways of disempowerment and conflict.

Fifth, the trusts serve as essential vehicles for implementing sustainable socio-economic development projects that, supposedly, contribute to the long-term sustainability and viability of communities. The notion of CDTs, however, has also been criticised for being somewhat of a two-edged mechanism: while it professes to represent community interests, and thereby presumably expands democratic ideals such as inclusiveness, responsiveness, and accountability, it simultaneously defends and promotes the business-centric interests of the mining companies. Finally, these development trusts, for years, breached their governing acts, which stipulate that they must be administered for the benefit and material welfare of the broader community.

In Weberian terms, it seems that a degree of elite capture (see the "Theoretical orientation – EGT and path dependency" section) has crept into many of these development trusts. An immediate consequence of this capture is that CDTs are often dominated by a few well-networked elites who claim to represent community interest. However, much remains to be known as to *why* CDTs have become enablers and primary conduits for the trust deficit between mining companies and communities in South Africa. One example that accentuates the outcome of this tension is the move towards agreement-making in mining, which is aimed at achieving acceptance and gaining approval from local communities. There has been a steady increase in the number of studies of CDAs; however, in the context of South Africa and elsewhere in the world, these agreements are not necessarily accompanied by an improvement in the overall well-being of mining host communities. I now turn to the concept of SLO.

The SLO in the mining industry

The scholarly discourse on the tacit social acceptance of mining companies by communities has been dominated by studies of SLO (Nelsen, 2006; Prno & Slocombe, 2012; Bice, 2014; Martineza & Franks, 2014; Mayes, 2015; Zhang et al., 2015), which gained momentum since 2002, after the International Institute of Environment and Development published the landmark report "Breaking New Ground: Mining, Minerals and Sustainable Development" (IIED, 2002). The origins of the concept can, however, be traced to the argument of Shocker and Prakash Sethi (1973), which enterprises need a social contract to operate. This social contract was primarily influenced by the behaviour of an enterprise, thereby creating an aspirational culture rooted in trust, respect, and engagement within a particular environment. Thus, a general concern of the government and civil society for the well-being of individuals led to more socially responsible business practices.

In the context of the extractive industry, early attempts to explain the SLO are linked to the increased perceptions of social risk, which highlighted the need for mining companies to build and cultivate solid relationships that encompass social accountability (Thomson & Boutilier, 2011). Much of this early literature on SLO draws heavily on the concepts of CSR (Thomson & Boutilier, 2011; Owen & Kemp, 2013), which has provided a constant frame of reference for later interpretations of social licensing. Since then, the evolution of SLO as a concept has had a seemingly complex pathway. For example, some notable features of an SLO are that it exists beyond any direct legal or governmental accountability (Joyce & Thomson, 2000). This feature confirms that, as a social construct, SLO is "a form of 'soft' regulation enforced through the beliefs and actions of relevant stakeholders" (Mercer-Mapstonea et al., 2017: 137) or "a discursive agreement between the actors involved without clear boundaries or requirements" (Meesters & Behagel, 2017: 275).

With respect to applied research, Moffat et al. (2015) argued that the concept has also been "examined to demonstrate how the roles of trust, fairness and governance may underpin the development of more sustainable, trust-based relationships between industry and society." More specifically, I use the concept of SLO as described by Thomson and Boutilier (2011) as the level of acceptance or approval continually granted to an organisation's operations or project by the local community and vested stakeholders.

Over the past decade, the scholarly discourse on SLO has been characterised by seemingly disparate thinking, leading to several distinct themes. One of the emerging themes is that SLO has a local dimension, that stakeholders are involved, and that it requires some form of consent. These themes have raised the profile of social concerns in the mining industry whose concerns are primarily technical (Owen & Kemp, 2013). A related common theme is that, theoretically, the conceptual roots of SLO are linked to sustainable development, CSR, and multilevel governance (Koivurova et al., 2015).

Several researchers define the term that comprises three normative components, namely, legitimacy, credibility, and trust (Thomson & Boutilier, 2011; Prno & Slocombe, 2012). Additionally, there are four levels of an SLO, namely withdrawal, acceptance, approval, and psychological identification, ranging from the lowest to highest, which represent "how the community treats the company" (Thomson & Boutilier, 2011: 1784). Moreover, according to Thomson and Boutilier (2011), trust is fundamental to moving through these four levels of SLO. In case of lack of trust, the so-called "social licence" may be in jeopardy (Bice, 2016), and low levels of community acceptance for mining projects pose significant risks to project success (Que et al., 2019).

Broadly, a multitude of factors may influence the distrust of mining companies, which include a lack of community engagement (Carstens & Hilson, 2009; Costanza, 2016); no immediate visible, tangible material benefits; unrealistic expectations; acquisition of land; and disruption to the social fabric of the local community (Bice, 2016). A recent study by Dagvadorj et al. (2018) identified three determinants of trust: motivation, ability, and effort. First, they explain trust as arising from the perception of fairness (for example, to what extent does a mining company take care of the issues facing the community). Second is the ability of the mining company to perform as a technically competent entity (competence-based trust). Third, the local community's perception of the efforts of a mining company to maintain environmental protection also affects how they trust such a company.

Another contribution to the theoretical insights about the central role of trust in shaping the outcome of the relations between mining companies and communities is made by Prno (2013) who assumes that mining companies can establish an SLO using five key actions, namely, local benefits provision, the building of relationships, an awareness of context, increased focus on the sustainability of operations, and the ability to adapt. Additionally, Thompson and Boutilier's (2011) cumulative pyramid model explains SLO by assuming a combination of three principal components: legitimacy, credibility, and trust. The emphasis in this model is that once a mining operation has developed legitimacy and credibility at a later stage, it is only then that acceptance and approval of the operation will follow. When a full trust relationship has been established, and the community and other interested and affected stakeholders become involved in the identification of needs, they will actively support the needs of the mining company (Boutilier & Thomson, 2011).

Following Thompson and Boutilier (2011), Moffat and Zhang's (2014) integrative model, explaining community acceptance of mining, offers a profound deliberation of how trust is central to mining companies to obtain and maintain an SLO. Drawing on social psychological research in intergroup relations, this model posits that trust is mainly dependent on three critical elements of a social licence, namely, the impact on social infrastructure, how companies engage with communities in respect of quantity and quality, and perceptions of procedural fairness. The model assumes that community trust in mining operations erodes when there is a negative impact of a mining operation on social infrastructure. Conversely, perceived procedural fairness and contact quality enhance trust. In turn, overall community trust will determine the extent to which community members accept or reject a mining operation (Moffat & Zhang, 2014).

Recently, however, there have been growing concerns about the practical application of the term "SLO." In particular, a significant concern for scholars, such as Bice (2014) and Owen and Kemp (2013), is the emphasis on the informal nature of the concept, as opposed to a formal contract (such as an environmental licence, for example). There is a growing tendency by the mining industry to view committees convened and controlled by mining companies as "best practice" in respect of SLO (Mayes, 2014). It is thus not surprising that the notion of SLO is portrayed as a barrier to discussion and debate at a local level (Owen & Kemp, 2013), and that the mining industry's risk management orientation limits its long-term planning (Owen & Kemp, 2014).

This ultimately leads to what has been conceptualised as an audit culture, which often contributes to consultation fatigue (Lacey, 2013) or what Owen (2016: 104) referred to as *Mineras Intteruptus*, an assumption inside the mining industry "that if disapproval becomes too intense there is a chance that members of the community will interrupt mining activities." A major challenge to the notion of SLO comes from Owen and Kemp (2017). In *Extractive Relations: Countervailing power and the global mining industry*, Owen and Kemp (2017: 50) note, "it is clear the idea of a social license is designed to privilege the 'majority' perspective" (Owen & Kemp, 2017: 50) and, thus, "can overlook established patterns of exclusion and even serve to reinforce them" (Ibid. 51).

The above divergent views offer a wide range of understanding of SLO. Nevertheless, there is much to be learned from these interpretations. I attempted to untangle the divergent views of the concept not to advance the debate about SLOs in this book. I do believe, however, that these views provide perspectives to advance some of the arguments in this book. A mining company adopts several ways to obtain a formal licence to operate in different parts of the world, and this concept can be applied to an SLO. In the context of this book, drawing from Mofatt and Zhang (2014), an SLO is granted when a community concludes and accepts the conditions of a CDA, which often stipulates the type of CSR projects to be implemented by a CDT.

Since the benefits promised by agreements do not flow automatically once they are signed, the relationship over time, the SLO is *maintained* by the trust relationship between the CDT or mine and community. The relationship is dependent on the positive impact on the social infrastructure (for example, the building of roads, schools, and health-care facilities), the quality of engagement, as well how community members feel about their treatment by a mining company or the CDT. Conversely, SLO is *threatened* by community mobilisation and *revoked* once these efforts by the community lead to the dismantlement of the CDT. Obviously, I do acknowledge that, in some instances, the pathways of obtaining and maintaining an SLO are far more complex than described here. Yet, at the same time, given the convergence of thought on the central role of trust in relations between a mining company and the community, this is an attempt to contribute to a more pragmatic application of SLO. Next, I provide a brief explanation of the term "governability," which is closely linked to the concept of social licensing.

Governability and SLO

Governance suggests that the market and civil society have prominent roles – along the state – in the governing of modern societies. Chuenpagdee et al. (2008: 3) define governability as "the overall capacity for governance of any societal entity or system." For example, it could refer to "the governance status of a societal sector or system such as a fishery or a coastal region as a whole" (Kooiman et al., 2008: 1). In this context, governability consists of three coherent analytical components: the system to be governed, its governing system, and their governance interactions (Kooiman et al., 2008). More specifically, governability

> provides a conceptual basis for assessing and improving the interactive governance of natural resource systems. There is a close relationship between the two concepts. An understanding that seeks to improve governance inevitably results in the need to explore and to assess governability. Governability of natural resource systems can vice versa only be understood by reference to their basic qualities.
>
> (Kooiman et al., 2008: 2)

Countries such as Switzerland have adopted participatory governance to increase governability. As a result, participatory governance serves as a means to enhance governability by providing "information to governments about the public acceptability of different policy options and strategies" (Kübler et al., 2019: 5). In the context of mining, the principles and values guiding governance of community-based natural-resource management play a central role in the acceptance of mining projects by mining-affected communities. Governance theory contains numerous examples of the significance of the modes of governance at the structural level: self-, co-, and hierarchical governance (Kooiman et al., 2008).

Theoretical orientation – EGT and path dependency

This section provides an outline of EGT and path dependency. The focus on governance spawned a substantial literature in critical disciplines, such as politics, economics, development studies, and socio-legal studies. The process of democratisation and globalisation (Chhotray & Stoker, 2009) has motivated the production of this literature. According to Chhotray and Stoker (2009: 2), "government seeks to know the way we construct collective decision-making."

Conversely, EGT understands governance as radically evolutionary, which implies that all elements of governance are subject to evolution, that these elements co-evolve, and that many of them are the product of governance itself (Van Assche et al., 2014; Beunen et al., 2016). Van Assche (2016: 20) defines governance as "taking of collectively binding decisions for a community in a community, by governmental and other actors." Within EGT, governance also includes various institutions, changing relationships, and governance paths. The "governance paths are histories of confrontations between these different versions of the world and different attempts to steer, govern and coordinate" (Van Assche, 2016: 209). EGT is seen as a novel contribution that allows us to understand changes in society and the interventions needed. Above all, it acknowledges the significance of including various actors beyond government in collective decision-making (Beunen et al., 2016).

With respect to path dependency, during the process of co-evolution, different evolutionary pathways that influence each other's development are created (Van Assche et al., 2014; Beunen et al., 2016). Path dependency helps in understanding how "legacies from the past shape future options" (Van Assche et al., 2016: 28) and explains how "successive generations of political and social actors have difficulty in departing from patterns set by their predecessors" (Crouch, 2010: 112). However, path dependency does not provide the means of predicting the future by analysing the past (North, 1990). In research, path dependency instead helps to explain and compare alternatives. The link between path dependency and institutions is worth discussing in more detail. Path dependencies are now closely linked to governance or institutional theories (Hall & Taylor, 1996). Proponents of path dependency often distinguish between three levels of institutions: the macro or constitutional level, the policy level, and the decision-making level (Kay, 2005).

Although, historically, most studies have focused on the macro level (Pierson, 1993), increasing attention is now also being devoted to the other two levels (Kemp, 2000). Path dependency is of value for policy analysis: it acknowledges that policy decisions accumulate over time, it emphasises the complexities associated with policy analysis, and it considers the policy system (and subsystems) as a whole (Kay, 2005). I shall return to path dependency with the evolutionary theory again in Chapter 2. Now, the question is why the focus is on social licensing?

Why social licensing and mining in South Africa?

Against the background of the theoretical underpinnings of the book in the previous section, I now turn to the rationale of the book. A community trust, as a form of CDA, is not a new phenomenon. However, the community trusts initiated by mining companies voluntarily have received little attention in scholarly contributions, with a few exceptions (ActionAid, 2008; ActionAidSA, 2016; Harvey, 2017; Matebesi & Marais, 2018, Leonard, 2019). Information about community trusts is mostly found in policy papers, websites, and annual reports of mining companies. The case of the Royal Bafokeng Development Trust (RBNDT) stands out as it has received widespread scholarly attention. Similarly, the widespread violent protests against community trusts under the leadership of traditional authorities have drawn little scholarly attention.

Social licensing and mining provides an indispensable source to understand – from local perspectives – the challenges mining communities face with community trusts, as well as the opportunities these trusts hold for the sustainable development of these areas. In particular, it is concerned with understanding *why* CDTs, which provide a crucial foundation upon which to implement CDAs in South Africa, have become part of the broader problems in the often-adversarial relations between mining companies and mining host communities.

For example, over the past two decades, CDTs have been the subject of high and rising levels of popular mobilisation by residents in mining host communities and labour-sending communities. What are the goals and objectives of CDAs? What is the governance structure of CDTs? How are members of the community being involved in the CDA decision-making and implementation processes? What measures are in place to ensure that projects initiated by the community trust contribute to a community and socio-economic development and sustainability? What are the challenges and opportunities for CDTs? Why are some successful and others not?

It is hoped that the book will make a distinctive contribution to the field of SLO in terms of three interrelated broad headings: community relations and development, sustainable development, and CSR. By demonstrating the determinative role played by community trusts in community protests, the author will challenge simplistic views of SLO that are often described as the primary source of costly conflict between mining companies and local communities. It seeks to expand the body of work on SLO and mining conflict in two primary ways. First, a great deal of empirical research speaks to the economic benefits of mining to local communities. This is crucial, but so are the effects of conflict on the social fabric of local communities. This measure receives considerable attention in this book.

Second, much of the recent research on mining conflict primarily focuses on mining companies, local opposition groups in mining communities, and the national state (Costanza, 2016). A crucial dimension that the book adds is the focus on path dependency within the EGT. The emphasis is on the relationship between the policy level and the decision-making levels in path dependency, which is an under-researched component of path dependency.

As reported elsewhere (Matebesi & Marais, 2018), path dependency enriches this book in three ways. First, it demonstrates how mining policies developed over decades continue to serve as a barrier to current mine-community and government engagements. Second, it converges the institutional analysis at the policy and the decision-making levels. Finally, we emphasise the confrontations between the different world views at the local level and how, despite structural change at the policy level, the existing conflict remains. The book will, therefore, be of interest not only to scholars in disciplines like sociology, political science, urban studies, and public management, but also to readers with an interest in mining–community relations in a developing country setting.

The evidence

This book is based on four case studies of community trusts in four provinces of South Africa: Itumeleng Community Trust (ICT) in Jagersfontein (Free State Province), Annooraq Community Participatory Trust Limpopo (Annooraq Trust) in Atok (Limpopo Province), JohnTaolo Gaetswe Development Trust (JTG Development Trust) in Kuruman (Northern Cape), and the Royal Bafokeng Nation Development Trust (RBNDT or RBN) in Rustenburg (North West Province) (Figure 1.1).

The community trusts were purposefully selected from different provinces in order to capture a broad spectrum of political and community patterns. Moreover, all four community trusts have experienced varying degrees of contestation, ranging from high (Atok and Kuruman) to moderate (Jagersfontein) and low (RBNDT). Based on my focus on the distinction between trusts, Atok CDT and RNBDT are managed by traditional authorities, whereas Jagersfontein and Kuruman community trusts are managed by a typical mining company, government, and civil society representatives. As shall be seen later, RNBDT is by far the most successful community trust at reaching meaningful benefits for local communities.

In order to answer the research questions posed earlier, the primary data were collected over three years (2016–2018). In this regard, semi-structured, in-depth, qualitative interviews were conducted with representatives of community trusts, MACUA-SA, command community leaders, the National Department of Mineral Resources, and members of a civil society organisation coalition formed by

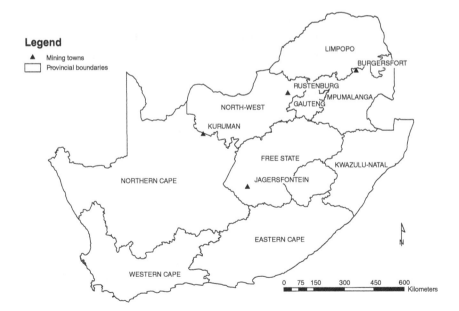

Figure 1.1 Map of South Africa and locations of case studies.

the MACUA-SA. Focus group sessions were also conducted with members of the community in the mining communities.

A survey of 200 randomly selected residents was also conducted in each of the four case study locations. The survey seeks to understand the knowledge of residents about the CDT, perceptions about community consultation and participation, local benefits, protest participation, as well as the changes they would like to see in the future. These issues were categorised as procedural fairness and trust in the CDT, perceptions about the contribution of the CDT on the quality of life, and future behavioural intentions.

Procedural fairness was measured with two questions: "I believe the CDT listen to and respect our opinions" and "I believe the CDT is prepared to change its practices in response to community sentiment." Trust was measured with three statements: "I can trust the CDT to act responsibly," "I trust the CDT to act in the best interest of the community," and "My trust in the CDT has decreased over the past five years." The following questions related to future intensions – measured with a seven-point Likert scale (1 = highly likely, 7 highly unlikely) – were asked:

- How likely will you complain to people about the CDT?
- How likely will you attempt to discourage others from working with CDT?
- How likely will you encourage others to protests against the CDT?
- How likely will you participate in a protest by residents against the CDT?
- How likely will your community be a better place to live in a decade from now?

Weaved together, the cases offer insights into the motivations and actions of diverse actors and the pathways in which they are locked due to the institutional arrangement and local decision-making processes.

Structure of the book

In dealing with the factors mentioned above, the book is organised into seven chapters – all informed by an extensive literature review of local and international case studies – and empirical findings of in-depth interviews and surveys. The first three chapters draw significantly on the literature on SLO and path dependency within EGT. Chapters 4 and 5 present the empirical findings, while Chapter 6 concludes.

The introductory chapter is followed by Chapter 2, which presents the theoretical framework underlying the book. After introducing the theoretical framework, the focus turns to reviewing previous research to identify the research gap. Additionally, the chapter presents the building blocks of the integrative model that explains community acceptance of mining. The four building blocks of this model, which determine the degree of trust, are the impact on social infrastructure, how companies engage with communities with respect to quantity and quality, and perceptions of procedural fairness. Afterwards, it turns to the elite capture theory that highlights the relations between different elites and how such interactions influence conflict between community trusts and local communities. Insights from these two theories are combined and used as a basis for interpreting the findings.

Chapter 3 focuses on the formal regulatory versus informal requirements to operate a mine in South Africa and beyond. In this chapter, the author seeks to contribute to current international debates on the regulatory requirements of two distinct yet interrelated dimensions of SLO, namely, formal (legal) and informal (social) licences. A government issues the former dimension in line with legislated requirements, and it endures as long as a company complies with the relevant conditions of the licence. Mining companies, however, need to earn a social licence from local communities and then continually renew and negotiate this in line with both the status and the quality of the relationship between the company and the community. The international literature suggests that SLO is a necessary but insufficient condition for acceptance and approval of mining. The central thesis is that the focus cannot only be on SLO but that variables must be placed in the context of the regulatory and socio-political realities of mining host communities. Against this background, the next part of the chapter traces the historical roots of mining regulatory requirements to the 1955 Freedom Charter and the MPRDA 28 of 2002. It looks specifically at the efforts to enhance the development of relations between mines and communities, thereby adding further depth to the background and the theoretical issues discussed in the first two chapters.

Chapters 4 and 5 present the empirical findings of the four case studies conducted in South Africa. Chapter 4 focuses on the presentation of the case of RBNDT. This analysis looks at why RBNDT has become successful to the extent that it is touted globally as a model of direct community participation in mining ventures. In particular, the chapter demonstrates how the Bafokeng Nation managed its shares from mining to transform from an agricultural community to a mining community with a vibrant and diverse local economy.

In Chapter 5, I turn to popular mobilisation against three community trusts. In some regions of South Africa – as is the case in Australia and Argentina, for example – community demands for relevant, direct and sustained benefits from mining have resulted in only a few sporadic popular mobilisations against mining companies and community trusts. In other regions, however, demands have led to sustained and quite violent popular mobilisation. This chapter sets out to examine this variation by comparing the case studies of two relatively small towns (Jagersfontein and Kuruman) and Atok, which is surrounded by several villages. Starting with a description of the political and economic conditions in the two towns, I move on to a more thorough investigation of community trusts and their link with popular mobilisation. In line with the path dependency within the EGT, the focus is on how the community actors are channelled along established policy paths and locked into pathways of conflict. Overall, the chapter points to the complexity faced by mining communities in asserting their rights in the decision-making processes that affect them.

18 *Introduction*

Chapter 6 provides a summary of the views by synthesising insights from previous chapters. It concludes that the existing SLO practices locally and internationally will undergo fundamental changes in years to come. The nature and direction of these changes will depend on the policy imperatives of the mining industry and on how community trusts engage with communities. The chapter ends with a discussion of the implications for contemporary mining company–community relations, in general, and for informal agreement-making processes and SLOs, in particular.

References

ActionAid (2008). Precious metals: The impact of Anglo Platinum on poor communities in Limpopo, South Africa [online]. Available at https://www.actionaid.org.uk/sites/default/files/doc_lib/angloplats_miningreport_aa.pdf (accessed 20 June 2017).

ActionAid SA (South Africa) (2016). Precious metals II: A previous inequality [online]. Available at https://south-africa.actionaid.org/publications/2016/precious-metals-ii-systemic-inequality (accessed 20 June 2017).

Adeleke, F. (2017). How South Africa's mining industry can change its ways. *The Conversation*, 1 March [online]. Available at https://theconversation.com/how-south-africas-mining-industry-can-change-its-ways-73214 (accessed 1 May 2018).

Bergek, A. & Onufrey, K. (2013). Is one path enough? Multiple paths and path interaction as an extension of path dependency theory. *Industrial and Corporate Change*, 23(5): 1261–1297.

Beunen, R., Van Assche, K. & Duineveld, M. (2016). *Evolutionary Governance Theory*. New York: Springer.

Bice, S. (2014). What gives you a social licence? An exploration of the social licence to operate in the Australian mining industry. *Resources*, 3(1): 62–80.

Bice, S. (2016). *Responsible mining: Key principles for industry integrity*. London: Routledge.

Boyd, W., Pruham, W.S. & Schurman, R.A. (2001). Industrial dynamics and the problem of nature. *Society and Natural Resources*, 14(7): 555–570.

Bunker, S. (1989). Staples, links, and poles in the construction of regional development theories. *Sociological Forum*, 4(4): 589–610.

Busacca, M. (2013). Corporate social responsibility in South Africa's mining industry: Redressing the legacy of apartheid. *CMC Senior Theses*, 632 [online]. Available at http://scholarship.claremont.edu/cmc_theses/632 (accessed 20 June 2017).

Brereton, D., Owen, J. & Kim, J. (2011). *Good practice note: Community development agreements*. EI Source Book. Centre for Energy, Petroleum and Mineral Law and Policy (CEPMLP), University of Dundee, Dundee.

Brosché, J. (2014). *Masters of war. The role of elites in Sudan's communal conflicts*. Report 102/Department of Peace and Conflict Research. Uppsala: Department of Peace and Conflict Research [online]. Available at https://uu.diva-portal.org/smash/get/diva2:677431/FULLTEXT01.pdf (accessed 20 June 2018).

Bruce, L. (2018). *Communities again excluded from Mining Charter discussions despite court order*. Centre for Applied Legal Studies, University of Witwatersrand, Johannesburg [online]. Available at www.wits.ac.za/news/sources/cals-news/2018/communities-again-excluded-from-mining-charter-discussions-despite-court-order.html (accessed 14 August 2018).

Carstens, J. & Hilson, G. (2009). Mining, grievance and conflict in rural Tanzania. *International Development Planning Review*, 31(3): 301–326.

Cawood, F.T. & Minnitt, R.C.A. (2001). A new royalty for South African mineral resources. *The Journal of the South African Institute of Mining and Metallurgy*, March/April: 91–96 [online]. Available at www.saimm.co.za/Journal/v101n02p091.pdf (accessed 20 June 2017).

Centre for Applied Legal Studies (2017). *The Social and Labours Plan series phase 2: Implementation operation analysis report*. Johannesburg: University of Witwatersrand.

Chabana, T. (2016). Presentation to SAHRC – National Hearing on underlying socio-economic challenges of mining-affected communities [online]. Available at www.mineralscouncil.org.za/industry-news/publications/presentations/send/7-2015/313-presentation-sahrc-national-hearing (accessed 20 June 2017).

Chhotray, V. & Stoker, G. (2009). *Governance theory and practice: A cross-discipline approach.* New York: Palgrave MacMillan.

Clapham, D. (2005). *The meaning of housing. A pathways approach.* Bristol: The Policy Press.

Conde, M. & Le Billon, P. (2017). Why do some communities resist mining projects while others do not? *Extractive Industries and Society*, 4(3): 681–697.

Costanza, J.N. (2016). Mining conflict and the politics of obtaining a Social License: Insight from Guatemala. *World Development*, 79(C): 97–113.

Dagvadorj, L., Byamba, B. & Ishikawa, M. (2018). Effect of local community's environmental perception on trust in a mining company: A case study in Mongolia. *Sustainability*, 10: 614.

Davenport, J. (2017). "Decolonising" South Africa's mining history. *Mining Weekly*, 14 April [online]. Available at www.miningweekly.com/login.php?url=/article/decolonising-south-africas-mining-history-2017-04-14 (accessed 14 September 2018).

Demmers, J. (2012). *Theories of violent conflict: An introduction*. London: Routledge.

De Beers Group (2009). Report to society [online]. Available at www.debeersgroup.com/~/media/Files/D/De-Beers-Group/documents/reports/library-/rts09-communities-may-2010.PDF (accessed April 2017).

Dlamini, T. & Mbangula, M. (2017). Where is the voice of mining-affected communities? *Business Day*, 20 June [online]. Available at www.businesslive.co.za/bd/opinion/2017-06-20-where-is-the-voice-of-mining-affected-communities/ (accessed 14 January 2018).

DMR (Department of Mineral Resources) (2017). *Broad-based black socio-economic empowerment charter for the South African mining and minerals industry*. Pretoria: Government Printer.

DMR (2018). *Broad-based socio-economic empowerment charter for the South African mining and minerals industry*. Pretoria: DMR.

Dougherty, M.L. & Olsen, T.D. (2014). They have good devices: Trust, mining, and the microsociology of environmental decision-making. *Journal of Cleaner Production*, 84(1): 183–192.

Etter-Phowa, R. (2017). How can Malawi ensure community development agreements are implemented effectively? *Mining Malawi*, 27 October [online]. Available at https://mininginmalawi.com/2017/10/27/how-can-malawi-ensure-community-development-agreements-are-implemented-effectively/ (accessed 14 January 2018).

Foucault, M. (1977). *Discipline and punish: The birth of the prison*. London: Allen Lane.

Franks, D. (2015). *Mountain movers: Mining, sustainability and the agents of change*. New York: Earthscan.

Fritzen, S.A. (2007). Can the design of community-driven development reduce the risk of elite capture? Evidence from Indonesia. *World Development*, 35(8): 1359–1375.

Godoy, E. (2013). Civil society seeks to influence Mexican mining law reform. *Inter Press Service*, 22 March [online]. Available at www.globalissues.org/news/2013/03/22/16154 (accessed 12 May 2016).

Groves, C., Munday, M. & Yakovleva, M. (2013). Fighting the pipe: Neoliberal governance and barriers to effective community participation in energy infrastructure planning. *Environment and Planning C. Politics and Space*, 31(2): 340–356.

Gunningham, N., Kagan, R. & Thornton, D. (2004). Social license and environmental protection: Why businesses go beyond compliance. *Law Social Inq*, 29(2): 307–341.

Harvey, R. (2017). Chartinga way for South African mining benefit communities. *SAIIA Policy Insights* no 50 (July) [online]. Available at www.africaportal.org/documents/17499/saia_spi_50_ harvey_20170808.pd (accessed 14 January 2018).

Hoffmann-Lange, U. (2007). Methods of elite research. In: R.R. Goodin (ed.). *The Oxford handbook of political behaviour* (pp. 910–927). Oxford: Oxford University Press.

Human Rights Watch. (2019). "We know our lives are in danger": Environment of Fear in South Africa's mining-affected communities [online]. Available at www.hrw.org/report/2019/04/16/we-know-our-lives-are-danger/environment-fear-south-africas-mining-affected (accessed 28 June 2019).

IIED (International Institute for Environment and Development). (2002). *Breaking new ground: Mining, minerals, and sustainable development*. London: Earthscan.

International Council on Mining and Mineral (ICMM). (2013). *Approaches to understanding development outcomes from mining*. Canberra: ICMM.

Jeffrey, A. (2018). *The 2018 Draft Mining Charter: Transformation trumps sustainability*. Johannesburg: South African Institute of Race Relations.

Joyce, S. & Thomson, I. (2000). Earning a social licence to operate: Social acceptability and resource development in Latin America. *Canadian Mining and Metallurgical Bulletin*, 93(1037): 49–52.

Kalyvas, S.N. (2003). The ontology of political violence. *Perspectives on Politics*, 1(3): 475–494.

Kalyvas, S.N. (2006). *The logic of violence in civil war*. Cambridge: Cambridge University Press.

Kay, A. (2005). A critique of the use of path dependency in policy studies. *Public Administration*. 83(3): 553–571.

Keenan, J.C., Kemp, D.L. & Ramsay, R.B. (2016). Company-community agreements, gender and development. *Journal of Business Ethics*, 135(4): 607–615.

Kemp, D. & Owen, J.R. (2013). Community relations and mining: Core to business but not "core business". *Resources Policy*, 38(4): 523–531.

Kemp, D. & Owen, J.R. (2015). Social science and the mining sector: Contemporary roles and dilemmas for engagement. In: S. Price & K.M. Robinson (eds). *Making a difference? Social assessment policy and praxis and its emergence in China* (pp. 60–83). New York: Berghahn Books.

Kemp, D., Owen, J.R., Gotzmann, N. & Bond, C.J. (2011). Just relations and company-community conflict in mining. *Journal of Business Ethics*, 101(1): 93–109.

Kirsch, S. (2014). *Mining capitalism: The relationship between corporations and their critics*. Berekeley: University of California Press.

Koivurova, T., Buanes, A., Riabova, L., Didyk, V., Ejdemo, T., Poelzer, G., Taavo, P. & Lesser, P. (2015). "Social license to operate": A relevant term in Northern European mining? *Polar Geography*, 38(3): 194–227.

Kooiman, J., Bavinck, M., Chuenpgdee, R., Mahon & Pullin, R. (2008). Interactive governance and governability: An Introduction. *The Journal of Transdisciplinary Environmental Studies*, 7(1): 1–11.

Kübler, D., Rochat, P.E., Woo, S.Y. & Van der Heiden, N. (2019). Strengthen governability rather than deepen democracy: Why local governments introduce participatory governance. *International Review of Administrative Sciences.*
Lacey, J. (2013). Can you legislate a social licence to operate? *The Conversation* [online]. Available at https://theconversation.com/can-you-legislate-a-social-licence-to-operate-10948 (accessed 14 January 2017).
Lariña, A.M., Dulce, J.C. & Saño, N. (2011). National and global energy governance: Issues and challenges in the Philippines. *Global Policy,* 2(S1): 80–93.
Leonard, L. (2017). State governance, participation and mining development: Lessons learned from Dullstroom, Mpumalanga. *Politikon,* 44(2): 327–345.
Leonard, L. (2019). Traditional leadership, community participation and mining development in South Africa: The case of Fuleni, Saint Lucia, KwaZulu-Natal. *Land Use Policy,* 86: 290–298.
López, M. (2013). Elite theory. *Sociopedia.isa.*
Loutit, L., Mandelbaum, J. & Szoke-Burke, S. (2016). Emerging practices in community development agreements. *Journal of Sustainable Development. Law & Policy,* 7(1): 64–96.
Lubbe, A. & Walaza, N. (2014). The pros and cons of community development trusts [online]. Available at http://tshikululu.org.za/the-pros-and-cons-of-community-development-trusts/ (accessed 14 January 2017).
Lund, J.F. & Saito-Jensen, M. (2013). Revisiting the issue of elite capture of participatory initiatives. *World Development,* 46: 104–112.
Malesa, G. & Jackson, N. (2017). South Africa. *The International Comparative Legal Guide to Mining Law 2018.* Chapter 28, pp. 196–203. London: ICGL.
Malope, L. (2018). Illegal mining with a conscience. *City Press Business,* 27 April, p. 5.
Marais, L. (2013). The impact of mine downscaling on the Free State Goldfields. *Urban Forum* 24(4): 503–521.
Marais, L., Nel, E. & Donaldson, R (eds). (2016). *Secondary cities and development.* London: Routledge.
Martineza, C. & Franks, D.M. (2014). Does mining company-sponsored community development influence social licence to operate? Evidence from private and state-owned companies in Chile. *Impact Assessment and Project Appraisal,* 32(4): 294–303.
Matebesi, S.Z. (2017). *Civil strife against local governance: Dynamics of community protests in contemporary South Africa.* London: Barbara Budrich.
Matebesi, S.Z. & Marais, L. (2018). Social licensing and mining in South Africa. *Resources Policy,* 59(1): 371–378.
Mayes, R. (2014). Mining and (sustainable) local communities: Transforming Ravensthorpe, Western Australia. In: M. Brueckner, A. Durey, R. Mayes & C. Pforr (eds). *Resource curse or cure? On the sustainability of development in Western Australia* (pp. 223–237). Heidelberg: Springer-Verlag.
Mayes, R. (2015). A social licence to operate: Corporate social responsibility, local communities and the constitution of global production networks. *Global Networks,* 15(s1): s109–s128.
Meesters, M.E. & Behagel, J.H. (2017). The social licence to operate: Ambiguities and the neutralization of harm in Mongolia. *Resources Policy,* 53(September): 274–282.
Mercer-Mapstonea, L., Rifkin, W., Moffat, K. & Louis, W. (2017). Conceptualising the role of dialogue in social licence to operate. *Resources Policy,* 54(2017): 136–146.
Modise, L.J. (2017). The notion of participatory democracy in relation to local ward committees: The distribution of power. *Skriflig,* 51(1): a2248.
Moffat, K. & Zhang, A. (2014). The paths to social licence to operate: An integrative model explaining community acceptance of mining. *Resources Policy,* 39(March): 61–70.

Moffat, K., Zhang, A., Lacey, J. & Leipold, S. (2015). The social licence to operate: A critical review. *Forestry*, 89(5).

Mosaval, R., Makwela, M. & Simelane, H. (2017). The intersection of participatory governance and Social Labour Plans in communities within mining towns: The experiences of Sikhululiwe village in Mpumalanga. Planact [online]. Available at www.planact.org.za/wp-content/uploads/2016/07/SoLG-2016-Planact-Contribution.pdf (accessed 3 August 2018).

Murambo, T. (2008). Beyond public participation: The disjuncture between South Africa's environmental impact assessment law and sustainable development. *PER: Potchefstroomse Elektroniese Regsblad*, 124–127 [online]. Available at www.scielo.org.za/scielo.php?-script=sci_arttext&pid=S1727-37812008000300006 (accessed 15 June 2017).

Musgrave, M.K. & Wong, S. (2016). Towards a more nuanced theory of elite capture in development projects. The importance of context and theories of power. *Journal of Sustainable Development*, 99(3): 87–103.

Nelsen, J.L. (2006). Social licence to operate. *International Journal of Mining, Reclamation and Environment*, 20(3): 161–162.

Nwapi, C. (2017). Legal and institutional frameworks for community development agreements in the mining sector in Africa. *The Extractive Industries and Society*, 4(1): 202–215.

O'Faircheallaigh, C. (2004). Evaluating agreements between indigenous peoples and resource developers. In: M. Langton, M. Tehan, L. Palmer & K. Shain (eds). *Honour among nations? Treaties and agreements with indigenous people* (pp. 303–328). Melbourne: Melbourne University Press.

O'Faircheallaigh, C. (2010). Aboriginal-mining company contractual agreements in Australia and Canada: Implications for political autonomy and community development. *Canadian Journal of Development Studies/Revue Canadienne d'études du développement*, 30(1–2): 69–86.

O'Faircheallaigh, C. (2011). Indigenous women and mining agreement negotiations: Australia and Canada. In: K. Lahiri-Dutt (ed). *Gendering the field: Towards sustainable livelihoods for mining communities* (pp. 87–110). Canberra: ANU E Press.

O'Faircheallaigh, C. (2012). Community development agreements in the mining industry: An emerging global phenomenon. *Community Development*, 44(2): 222–238.

Owen, J.R. (2016). Social license and the fear of Mineras Interruptus. *Geoforum*, 77: 102–105.

Owen, J. & Kemp, D. (2013). Social licence and mining: A critical perspective. *Resources Policy*, 38(1): 29–35.

Owen, J. & Kemp, D. (2014). Mining and community relations: Mapping the internal dimensions of practice. *Extractive Industries and Society*, 1(1): 12–19.

Owen, J. & Kemp, D. (2017). *Extractive relations: Countervailing power and the global mining industry*. New York: Routledge.

Parasuraman, A., Zeithaml, V.A. & Berry, L.L. (1985). Conceptual model of service quality and its implications for future research. *Journal of Marketing*, 49: 41–50.

Piper, L. & Nadvi, L. (2010). Popular mobilisation, party dominance and participatory governance in South Africa. In: L. Thompson & C. Tapscott (eds). *Citizenship and social movements: Perspectives from the global South* (pp. 212–238). London: Zed Books.

Pollis, A. (1996). Cultural relativism revisited: Through a state prism. *Human Rights Quarterly*, 18(2): 316–344.

Prno, J. & Scott Slocombe, D. (2012). Exploring the origins of "social license to operate" in the mining sector: Perspectives from governance and sustainability theories. *Resources Policy*, 37(3): 346–357.

Prno, J. (2013). An analysis of factors leading to the establish of a social licence to operate in the mining industry. *Resources Policy*, 38(4): 577–590.

Que, S., Wang, S., Awuah-Offei, K., Yang, W. & Jiang, H. (2019). Corporate social responsibility: Understanding the mining stakeholder with a case study. *Sustainability*, 11: 2407.

Ramatji, K.N. (2013). A legal analysis of the Mineral and Petroleum Resources Development Act (MPRDA) 28 of 2002 and its impact in the mining operations in the Limpopo Province [online]. Available at http://ulspace.ul.ac.za/bitstream/handle/10386/979/ramatji_kn_2013.pdf?sequence=1 (accessed 14 January 2017).

Rametse, M.S. (2016). *The significance of the Freedom Charter in the ideological debates within the ruling ANC Alliance in South Africa*. African Studies Association of Australasia and the Pacific (AFSAAP) Proceedings of the 38th AFSAAP Conference: 21st Century Tensions and Transformation in Africa, Deakin University, 28th–30th October, 2015 [online]. Available at http://afsaap.org.au/assets/Mochekoe_Stephen_Rametse_AFSAAP2015.pdf (accessed 16 April 2017).

Reis, E.P. & Moore, M. (2005). Elites, perceptions and poverties. In: E.P. Reis & M. Moore (eds). *Elite perceptions of poverty and inequality* (pp. 1–25). London: Zed Books.

Ribot, J.C. (2006). Choose democracy: Environmentalists socio-political responsibility. *Global Environmental Change*, 16(2): 115–119.

Rutledge, C. (2018). The mining charter: Citizens or subjects? *Business Day*, 16 February [online]. Available at www.businesslive.co.za-/bd/opinion/2018-02-16-the-mining-charter-citizens-or-subjects/ (accessed 15 April 2019).

Sarkar, S., Gow-Smith, A., Morakinyo, T., Frau, R. & Kuniholm, M. (2010). *Mining community development agreements – Practical experiences and field studies*. Washington: World Bank.

Shocker, A. & S. Prakash Sethi. (1973). An approach to incorporating societal preferences in developing corporate action strategies. *California Management Review*, 15(4): 97–105.

Sidley, K. (2015). Are community trusts the best approach to development? [online]. Available at http://trialogue.co.za/are-community-trusts-the-best-approach-to-development/ (accessed 16 April 2017).

Suttner, R. & Cronin, J. (1986). 30 years of the freedom charter. *Transformation*, 6: 73–86.

Tapscott, C. (2017). South Africa in the twenty-first century: Governance challenges in the struggle for social equity and economic growth. *Chinese Political Science Review*, 2(1): 69–84.

Terwindt, C. & Schliemann, C. (2017). *Tricky business: Space for civil society in natural resource struggles. Publication series on democracy*. Berlin: Heinrich Böll Foundation or the European Center for Constitutional and Human Rights [online]. Available at www.boell.de/sites/default/files/tricky-business.pdf (accessed 27 May 2018).

Thomson, I. & Boutilier, R.G. (2011). The social license to operate. In: P. Darling (ed.). *SME mining engineering handbook* (3rd edn, pp. 1779–1790). Littleton: Society for Mining, Metallurgy, and Exploration.

Thompson, I., Tapscott, C. & De Wet, P.T. (2018). An exploration of the concept of community and its impact on participatory governance policy and service delivery in poor areas of Cape Town, South Africa. *Politikon*, 45(2): 276–290.

United Nations Economic Commission for Africa (UNECA) (2011). *Minerals and Africa's development: The International Study Group Report on Africa's mineral regimes*. Ethiopia: UNECA.

Weber, M. (2005). *Economia y sociedad*. Mexico: Fondo de Cultura Económica.

World Bank. (2012). *Mining community development agreements source book* [online]. Available at http://siteresources.worldbank.org/INTOGMC/Resources/mining_community.pdf. (accessed 20 June 2017).

Xavier, R., Komendantova, N., Jarbandhan, V. & Nel, D. (2017). Participatory governance in the transformation of the South African energy sector: Critical success factors for environmental leadership. *Journal of Cleaner Production*, 15(June): 621–632.

Van Assche, K., Beunen, R. & Duineveld, M. (2014). *Evolutionary governance theory: An introduction*. Heidelberg: Springer.

Van Assche, K., Beunen, R. & Duineveld, M. (2016). An overview of EGT's main concepts. In: Beunen, R., Van Assche, K. & Duineveld, M. (eds). *Evolutionary governance theory* (pp. 19–36). New York: Springer.

Yamokoski, A. & Dubrow, J.K. (2008). How do elites define influence? Personality and respect as sources of social power. *Sociological Focus*, 41(4): 319–336.

Zhang, A., Moffat, K., Lacey, J., Wang, J., González, R., Uribe, K., Cui, L. & Dai, Y. (2015). Understanding the social licence to operate of mining at the national scale: A comparative study of Australia, China and Chile. *Journal of Cleaner Production*, 108(Part A): 1063–1072.

2 Path dependency and social licensing

Introduction

The primary aim of this chapter is to develop a theoretical explanation of the conditions under which, and the mechanisms through which mining companies obtain a social licence to operate (SLO). An SLO refers to community acceptance of large-scale projects like mining (Joyce & Thompson, 2000; Thompson & Boutilier, 2011; Franks & Cohen, 2012; Prno & Slocombe, 2012; Moffat & Zhang, 2014). A social licence highlights the importance of relationships, which are often underpinned by the notion of trust. Trust has been analysed in different fields including economics, psychology, and sociology. "Economic approaches viewed trust as either calculative or institutional and operationalised, for example, as the willingness to make wealth-creating investments without the guarantee of personal gain" (Maccannon et al., 2017: 2). Psychologists study trusting behaviour regarding underlying dispositions, intergroup processes, and cognitive expectations (Evans & Krueger, 2009).

For sociologists, trust is a central feature of social exchange of relations among people or institutions. Granovetter (1985), for example, argues that most behaviour is closely embedded in networks of interpersonal relations and that social relations, rather than institutional arrangements or generalised morality, are mainly responsible for the production of trust in economic relations. He further emphasises that while social relations may indeed often be a necessary condition for trust and trustworthy behaviour, they are not sufficient to guarantee these and may even provide occasion and means for conflict. Granovetter (1985) provides three reasons for this occurrence. First is that trust prompted by personal relations presents enhanced opportunity for malfeasance. Second, force and fraud are most efficiently pursued by teams, and the structure of these teams requires a level of internal trust. Third, the extent of disorder resulting from force and fraud depends very much on how the network of social relations is structured "where conflicts are relatively tame unless each side can escalate by calling on substantial numbers of allies in other firms, as sometimes happens in attempts to implement or forestall takeovers" (Granovetter, 1985: 493).

Rosenburg (1956) is credited with coining the first known systematic measurement of trust, referred to as the "Most-People" construct: "Some people say most people can be trusted. Others say you can't be too careful in your dealings with people. How do you feel about it?" (Rosenburg, 1956: 650). Since then, trust has become a concept in social sciences closely associated with several different studies focusing on, for example, civic culture (1963), political trust (Stokes, 1962; Fitzgerald & Wolak, 2016), and even social capital (Putnam, 2000). A current dilemma for the study of trust is that its conceptualisation is as divergent as ever. Despite the persistence of divergent views as to its meaning, theoretical underpinning, and measurement, the study of trust is even more relevant given today's global environment characterised by complex relationships at all levels. This relevance is underscored by significant recent contributions on the topic, including *The Oxford Handbook of Social and Political Trust* (Uslaner, 2018).

In the extant body of literature on SLO, the notions of social acceptance and public trust in mining companies are gaining traction (Joyce & Thompson, 2000; Boutilier & Thompson, 2011; Prno & Slocombe, 2012; Koivurova et al., 2013; Moffat & Zhang, 2014; Parsons & Moffat, 2014). This growth in the prominence of the concept of an SLO in extractive industries has drawn criticism, particularly regarding its pragmatic applicability (Owen & Kemp, 2013; Bice, 2014; Mayes, 2014; Owen & Kemp, 2014, 2017) with a need for a more systematic and nuanced terminological use (Gehman et al., 2017). It has become increasingly apparent that an SLO can be applied in some project contexts (Boutlier et al., 2012; Costanza, 2016; Meesters & Behagel, 2017) or local agreement-making contexts.

This chapter outlines the usefulness of applying Moffat and Zhang's (2014) integrative model in explaining the community acceptance of mining to studies of the role of trust in the mechanisms and processes of an SLO. In particular, it focuses on identifying the *actors* and *material processes* in the acquisition, maintenance, and diminishment involved in an SLO in community-based agreement-making contexts. Furthermore, I use insights from elite capture theories to frame the analysis of power relations within community-based implementing units for corporate social responsibility (CSR) or benefit sharing. However, of even greater importance, the chapter incorporates insights from these diverse theories to the path dependency within EGT. In this regard, three ways of approaching path dependence, as identified by Layton and Duffy (2018), are discussed.

The multiple underlying assumptions of trust

There are several theoretical debates about the term "trust." Rather than entering the different theoretical debates, the focus is on the socially embedded elements of trust, which are likely to present themselves in the case of agreement-making between companies and communities in a social (mining) context. A few of the divergent views on trust, however, need some consideration.

In *The Moral Basis of a Backward Society*, Banfield (1958) was interested in the mechanisms driving the generation of collective action and the increase in generalised trust. Using deterministic arguments, he constructed his analysis of "amoral familism" and "lower-class present orientations," and argued that there is something embedded in culture that makes people behave in a certain way, with which public intervention cannot deal directly (Ferragina, 2009). Amoral familism is generated by socio-economic circumstances (leading to a high death rate), historical circumstances (resulting in certain conditions of land tenure), and purely cultural circumstances, namely the absence of the institution of the extended family. According to Banfield, the lack of trust and collective action includes, for example, the absence of local newspapers or the strong control of the central administration.

While Fukuyama (1995) presents a convincing argument about the central role of trust and "spontaneous sociability" in business success and economic prosperity, Tilly (2004: 4) argues that trust "consists of placing valued outcomes at risk to others' malfeasance." For Fukuyama (1995), there are different pathways to trust. He posits that relationships depend on trust, on the one hand. Trust, on the other hand, depends on a culture of shared values. He viewed the rapid increase in legal representatives and prisoners, the breakdown of families, and voluntary community associations as signal of a decline in trust. Trust relationships, reports Tilly (2004), "include those in which people regularly take such risks." These trust relationships operate within larger networks (trust networks) that consist of "ramified interpersonal connections within which people set valued consequential, long-term resources and enterprises at risk to the malfeasance of others" (ibid. 5).

Further divergent views are offered by Uslaner (2018), for example, who somewhat agrees with Hardin's (1998) and Newton et al.'s (2018) notion that trust can be seen as a three-way relationship (A trust B to do X). He noted two concerns with this three-way relationship approach. First, it fails to resolve collective action problems. Second, it does not "help us bridge with people who are unlike us" (Uslaner, 2018: 6). In fact, in an attempt to complement Hardin's three-way construct of trust, Ripperger (1998) provides a better understanding of the trust relationship, including certain assumptions about psychological states (Zanini & Migueles, 2013). In doing so, Ripperger encapsulated the emotional and cognitive components of trust that coexist, but one can also prevail over the other depending on the specific situation and the people with whom we interact. It is therefore assumed that the motivations of people are highly dependent on the emotions or cognition that is influenced by the particular person or group of persons, situation, contexts, and circumstances (Zanini & Migueles, 2013).

Uslaner (2018) – who has written extensively on trust – bridges these different insights in the social sciences by distinguishing between generalised trust (social trust), particularised trust, and political trust. He maintained that generalised trust is the belief that most people can be trusted and is rooted in the process of socialisation. This generalised trust, he suggested, rests upon the psychological foundation of optimism and control, conceptualised as: "The future looks bright, and I can help make it better" (ibid. 4). Several notable elements of generalised can be observed. First, generalised trust is general and not related to anyone in particular (Igarashi et al., 2008). Second, it is stable over time and, third, negatively related to confidence in political institutions (Uslaner, 2018). Fourth, generalised trust is also associated with an ability to obtain correct information from others (Kikuchi et al., 1997). Conversely, particularised trust refers to "faith only in people like yourself" (Uslaner, 2018: 4), and is based on individuals' ties to their own in-groups (Harvey, 2014; Uslaner, 2018). Political trust, on the other hand, denotes political confidence in institutions (the executive, the legislature, and the police, among others), and is based on evaluations about their performance (Uslaner, 2018). The literature on political trust has identified a range of potential factors underpinning trust, such as citizen satisfaction with policy, economic performance, the prevalence of political scandals and corruption, and the influence of social capital (Parker et al., 2014; Cole et al., 2018).

In line with the approach of this book, I agree with Rousseau et al. (1998) that trust is neither a behaviour nor a choice, but an underlying psychological belief that allows the trustor to trust the trustee. Judging trust in this manner highlights the importance of the trust context. Adali (2013: 11) aptly describes the trust context as a system-level description of trust evaluation that takes as input four elements: first, the trustor and the network she is operating in; second, her goal(s) in making a trust evaluation and the underlying trust constructs; third, the trustee(s) that she depends on; and fourth, the environmental conditions that impact the trust evaluation. However, despite all the extant theoretical variants of trust, Cohen's (2016) review of *Public Trust in Business* (edited by Harris et al., 2016) highlighted a fundamental flaw in some of the literature: technical difficulty with the conceptualisations of trust.

The concept of trust is an important one in the literature on legitimacy and can be seen as central to all decision-making procedures (De Jong & Humphreys, 2016), as well as those that relate to an SLO (Thompson & Boutilier, 2011). In this context, trust is defined as the reliance of one actor on the truth, honesty, and integrity of another (De Jong & Humphreys, 2016). The authors argued that the concept of an SLO needs to be understood in the context of neoliberal environmental governance, which is the political view that the collective good is best realised when a free market system is adopted. A salient feature of such an understanding of neoliberal environmental governance is that the primary role of the state should be confined to providing the enabling conditions for trade, economic development, and private-sector investment (ibid.). In this conception, where the trustor is the host community of a mining operation and the trustee refers to the mining company, a loss of legitimacy and the withdrawal of the SLO are dependent on three kinds of trust: integrity-based trust, competency-based trust (Poppo & Schepker, 2010; Moffat & Zhang, 2014), and institutionalised trust (2017).

According to Moffat and Zhang (2014), when integrity- and competency-based trusts are lacking, the quality of the relationship between the community and the mining company is likely to deteriorate. Integrity-based trust refers to the trust that is created when the trustor believes that the trustee adheres to certain principles, whereas competency-based trust is based on the trustor's view of the skills and knowledge of the trustee about the work of the latter. Another form of trust, which is stronger than the former two, is institutionalised trust, which is fostered by regular interaction between the trustor and the trustee (Black, 2013). Depending on the influence of other contextual factors, including elite actors, institutionalised trust is more likely to contribute to a trust relationship between a community and a mining company. As Rousseau et al. (1998) correctly note, such trust not only has a strong positive influence on the strength of the relationship but may also yield positive project outcomes.

Following Cohen's (2016) concern and drawing on the vast literature on mining company–community relations, I describe the obligations of mining companies towards communities and stakeholders to include the protection and provision of economic and social benefits. I adopted Cohen and Dienhart's (2013: 4) definition of trust:

> When A trusts B to do X, A invites B to acknowledge and accept an obligation to do X. When – or if – B accepts the invitation, B takes on that obligation. In that way, trust creates an obligation and forms a trust relationship.

This account of trust highlights its proper characterisation and, in the context of this book, refers to how a community trusts that a mining company will engage sustainable mining practices and provide economic, social, and other benefits, as identified in informal agreements in the contexts of community developments trusts.

Community trust and the SLO – a pathway approach

The concept of SLO has received much criticism, in particular being "idealistic," "intangible," and difficult to measure (Owen & Kemp, 2012). In *Extractive Relations*, Owen and Kemp (2017) perhaps provide the most profound critique of the term "social licence":

> Our principal objection to the term "social licence" is its "inflated" status and what its usage infers. Despite industry acknowledgement that a social licence is not materially available to companies in ways that are comparable to a regulatory licence for exploration or mining, industry usage rests on an assumption that a "social licence" can indeed be granted by the communities or stakeholders their operations impact upon. In contrast, legally mandated instruments have specified conditions and obligations, which are monitored by a regulatory authority typically with power to revoke the licence should conditions not be met.
>
> Owen & Kemp, 2017: 41

While noting these criticisms against the term "social licence," I avoid rehearsing the other main criticisms against the term here. However, developing alternative terminology to the term – Owen and Kemp (2018) do not explicitly advocate this – will only express some of the same concerns about it. There is evidence, though, "that the drivers of SLO at the local operational level … can be systematically modelled and measured by conducting large-scale surveys of citizen attitudes" (Zhang et al., 2015: 484). The text now turns to the model of SLO, delineating how mining companies may transit a pathway towards establishing community acceptance and approval of a mining operation by enhancing the perception of impact from mining operations, intergroup contact experiences, and the perception of procedural fairness (Moffat & Zhang, 2014). This model is juxtaposed with current research.

Impact on social infrastructure

The integrative model explaining community acceptance and approval of mining postulates that the negative impact of a mining operation on the social infrastructure reduces community trust in the operation. This acceptance, Moffat and Zhang (2014) argued, depends on how residents have experienced the impact, compared to what they initially expected. In this approach, the emphasis is on the community's perceptions of the overall impact: If it is worse than expected, it may presumably erode trust in the mining company and eventually lead to the rejection of the mining development in that particular area. Conversely, therefore, it implies that if the net effect on the social infrastructure is perceived as being positive, it cultivates community trust, which substantially increases the possibility of community acceptance of the mining operations (ibid.).

A considerable body of literature deals with the direct and indirect links between mining and the impact on the social infrastructure (Hajkowicz et al., 2011; O'Faircheallaigh, 2011; Kemp & Owen, 2013; Martineza & Franks, 2014; Mayes, 2014; Wesley & MacCallum, 2014; Maluleke & Pretorius, 2016). About the importance of measuring the broad range of potential social impacts, Brueckner et al. (2014: 221) aptly noted that development-related social impacts are often "masked by economic assumptions made about, and indicators produced of, the employment and income potential development is believed to deliver." These factors are often portrayed to be outweighing the potentially adverse social consequences of economic pursuits, proving difficult the task of challenging the dominant narrative (ibid. 221).

Research generally supports the idea that where mining construction is positively associated with household income, housing affordability, access to household income, access to communication services, educational attainment, increased life expectancy, and trust of the mining company will increase substantially (Mayes, 2014; Maluleke & Pretorius, 2016). Several studies in South Africa highlight how mining companies invest in various social development projects, including sport and tourism infrastructure at a community level (Economix Research, 2016). This further includes a focus on early childhood development centres; mathematics, science, and language development at school level; strengthening the public health-care system; and capacity building for community-based and other non-governmental organisations by the Anglo American Chairman's Fund (the most significant corporate donor in the country) at a broader societal level (Federation for a Sustainable Environment, 2018). There is evidence that some mining companies invest in social infrastructure outside of their immediate sphere of operations. In the low-income and remote communities of Guatemala, for example, some gold mining companies are working to help overcome challenges such as access to financial products faced by residents. The local community now has access to a bank branch and two automated teller machines (World Gold Council, 2018). According to Moffat and Zhang (2014), if there is convergence between the expected and experienced impacts, it may positively affect the extent to which community members accept the mining operation.

In most parts of the world, the government regulates the formal licence to operate (Transparency International, 2017). Such measures often lead to (relatively) better benefits for local communities and may create the situation in which business-centric interests and local community needs and expectations are not competing outcomes of mining. The study of Mayes (2014) on the local perspectives on mining in Ravensthorpe, Australia, however, demonstrates the disastrous consequences of mining with little sustainable responsibility or commitment beyond business imperatives. What transpired in Ravensthorpe was that BHP Billiton repeated public promises that the local mine would have a 25-year life span. Meanwhile, the community endured ongoing restructuring of community identity and sense of place, only to be met with the sudden closure of the mine after only eight months of operation (ibid.).

While this highlights the ambiguity inherent in mining operations, the distrust such an unexpected event may invoke among residents highlights two concerns, among others: It increases the likelihood for future mining developments to be rejected by the community, and as Sarkar et al. (2010) noted, mining activities or community development agreements must fit into the context surrounding sustainability.

Chuhan-Pole et al. (2017) suggested that in general, the socio-economic effects of large-scale mining are not well understood, but public opinion on the impact of mining on local communities is likely to be unfavourable. They argued that the unfavourable opinion of the public, in part, is because the social impact generated by mines is generally modest. In this context, it implies that a mining company will find it challenging to obtain community acceptance. There are, however, some problems with this notion of unfavourable public opinion on the social impact of mining. Indeed, not many local communities will possibly know or – when they do know – understand the macroeconomic booms of mining, including export revenues and gross domestic product. Similarly, a community will not know about skills transfer to residents as a result of their employment in the mine or the multiplier effect on the growth of small and emerging business enterprises, unless informed by someone. The literature, however, is replete with examples of the substantial social impact of mining at the local community level which has created positive relationships between mining companies and communities (for example, see Mancini & Sala's, 2018, extensive review).

Contact between the local community and the mining company

An underlying notion of Moffat and Zhang's (2014) integrative model is the fact that positive contact between the mining company personnel and community members should stimulate goodwill and trust, which will subsequently increase the likelihood that mining developments will be accepted by a community, and, in turn, that a social licence will be granted. It is well supported that intergroup contact reduces prejudice and that positive contact can increase trust and improve attitudes between groups (McKeown & Psaltis, 2017).

Some scholars contend that in countries where distrust is universal, the early establishment of good relations between mining companies and stakeholders is essential (Kempt et al., 2011). Others demonstrated that the benefits of intergroup contact are broader than previously thought, in that contact not only changes attitudes for individuals experiencing direct positive intergroup contact, but "their attitudes are also influenced by the behaviour (and norms) of fellow ingroup members in their social context" (Christ et al., 2014: 3996). In the context of mining developments or community development agreement-making, communities may feel positive about the potential for mutual benefit (based on trust) from mining operations (Thompson & Boutilier, 2011) and may enhance a mining company's chance to obtain a social licence.

While Moffat and Zhang (2014) emphasised the importance for a mining company to establish positive contact, and hence trust, with a community, in reality, however, there are several outcomes of contact. The majority of research to date, however, has focused on the outcomes of positive intergroup contact (Hewstone, 2009; Pettigrew & Tropp, 2011; Christ et al., 2014; Pettigrew & Hewstone, 2017), which is often accompanied by beneficial outcomes for an interaction, including cooperative interactions with positive joint outcomes, learning, or feelings of appreciation (Kauff et al., 2017). Other streams of research suggest a strong association between harmful contact and increased prejudice (Barlow et al., 2012; Kauff et al., 2017). Harmful contact is characterised by negative outcomes, including competitive interactions with negative outcomes for one of the interaction partners, hostile behaviour by the interaction partner, or feelings of being exploited (Hewstone et al., 2017). Reimer et al. (2017) drew our attention to the importance of considering the potential mobilising effect of negative contact on collective action. Negative contact raises the salience of group membership (i.e., disadvantaged-group members to think and feel as group members rather than individuals) and could thus evoke perceptions of discrimination (Wright, 2013), as well as inciting group-based anger (Reimer et al., 2017).

How these notions of intergroup unfold can be further understood by considering the broader context of mining company–community relations in South Africa, and contact quality. Mining in South Africa represents spaces where historical patterns of racial hierarchy (Lelyveld, 1981) are characterised by low-skilled migrant workers staying in informal settlements (Federation for a Sustainable Environment, 2018). Employment takes place in the context of personal and political relations (Jinnah, 2017), and is an important contextual factor that helps to understand not only workers' responses to employment in mines, but also how it shapes community opinions about mining companies. Using the Marikana massacre as a reference point, it is contended that the killing of 34 and injuring of 78 striking mineworkers by members of the South African Police Service (Bond & Mottiar, 2013; Duncan, 2014), beyond the spaces negotiated and resisted, could encourage disadvantaged-group members (in this case black communities in South Africa) to think and feel as group members rather than individuals. The collective identity of the workers, in turn, may have led to heightened levels of distrust of mining companies in South Africa.

There are a plethora of avenues for the mining sector to engage with the government development-related issues in South Africa. These avenues include the Social and Labour Plan (SLP) framework and the Mining Charter, an established instrument that provides a space for mining to talk to each other about their respective efforts in pursuing development partnerships. What lacks are platforms to engage with communities. In an era of declining trust in mining companies across most nations, the philosophy by some mining companies to voluntarily mobilise corporate equity ownership through community-based implementing structures, such as community development trusts (CDTs), in the context of South Africa (Benton, 2018), may provide opportunities for trust building between mining companies and communities.

For example, as articulated by Benton (2018), the process of establishing CDTs, which involves the selection of initial trustees, follows years of engagement between a mining company and community members assisting with the establishment of the trust. While the process of establishing CDTs is complicated and onerous (ibid.), studies of positive engagement have demonstrated how long-term effects of positive contact can enhance trust between interacting groups. This capacity to build trust is not realised by all mining companies in South Africa. Many mining companies devote the necessary resources to ensuring that CDTs are established, but they fail to devote comparable resources to ensuring that the CDTs, once established, are managed for the benefit of the entire community.

Procedural fairness

Procedural fairness is defined as the extent to which individuals perceive that they had reasonable involvement in a decision-making process (Tyler, 2006; Herian et al., 2012). Generally, though, the literature reveals that there are two interrelated ways of ensuring the perception of procedural justice. First is the recognition of being listened (Tyler, 2006; Herian et al., 2012; Lacey et al., 2016). In the context of SLO in the mining industry, Moffat and Zhang (2014), as well as several other studies (Thompson & Boutilier, 2011; Koivurova, 2015; Constanza, 2016; Moffat et al., 2016; Meesters & Behagel, 2017), have shown that the inclusion of public input can increase perceptions of procedural fairness among the public. Second is the perception of being treated with respect and dignity (Thompson & Boutilier, 2011; Williams & Walton, 2013). The perception of fair treatment is one of the strongest predictors of trust, and as Moffat and Zhang (2014) assumed, community members are more likely to accept mining operations in their region if mining companies are perceived as being fair in their decision-making processes.

Public participation in a myriad of government participatory forums and processes is a central feature in most democratic countries (Piper & Nadvi, 2010; Terwindt & Schliemann, 2017; Matebesi, 2017; Thompson et al., 2018).

Research has long suggested a link between established public participation and public trust in authority (Melkote & Steeves, 2015), and is often described as community engagement at a grassroots level. A positive or negative culture of community engagement is reflected in the attitude and behaviour of companies. For some companies, engagement has become an integral part of their business and, thus, is the best way to mitigate these community-related risks and achieve sustainable outcomes like an SLO, for example (Wang et al., 2016).

Other companies, however, see community engagement as being separate from business, or even too costly. Such companies, more than often, have seen their projects postponed, interrupted, and even shut down due to poor community engagement (Browne et al., 2011; Thompson & Boutilier, 2011). A study of the Waihi gold mining operations in New Zealand provides a practical example of how community engagement tends to foster mutual understanding and trust, leading to higher levels of acceptance of those operations. The Newmont Waihi Gold Company's announcement of its intention to extend its operations beneath homes in the residential area of Waihi in 2011 was subjected to an extensive three-year consultation period with the community to negotiate whether and how the extension might proceed (Lacey et al., 2016).

In some countries, there are legal requirements for community engagement. In South Africa, for example, the Mineral Resources and Petroleum Development Act of 2002 requires applicants to submit SLPs to the Department of Mineral Resources (DMR) for approval before the mining rights be granted. Broadly, the SLPs outline the company's commitment to contribute to local community development, but often the government grants mining rights without the consent of communities. Similarly, SLPs are mostly formulated between mining companies, political leaders, and traditional leaders, without the involvement of communities (Federation for a Sustainable Environment, 2018). Such events, which should have a high influence on trust and community acceptance of mining operations, are compounded in several ways. First is the dual and conflicting mandate of the DMR: to promote and regulate mining. For example, the biases of the DMR were perhaps more evident when the former minister of the department noted that mining must happen wherever there are minerals to be mined (Merten, 2016).

Second is when CDTs fall within the jurisdiction of traditional leaders. These leaders are often accused of trading the rights of communities for personal gain from mining companies (Federation for a Sustainable Environment, 2018). For example, in what is not an isolated case in South Africa, the *Daily Maverick* (2016) reported that the Mogalakwena platinum mine run by Anglo American Platinum (Amplats) in Limpopo, South Africa, supposedly gave 175 million ZAR (South African currency) to the Mapela Trust under the leadership of the tribal authority. Members of the community confronted the mine stating that they know nothing about the agreement between the mining company and the traditional leaders, the trust or the cash (ibid.). To date, the mine has been closed several times as a result of community protests over the lack of transparency and accountability of the money paid over to the Mapela Trust by Amplats (Federation for a Sustainable Environment, 2018). The protest against the Mogalakwena platinum mine not only highlights the blurred boundaries between the mining operations of the company and CDTs but also demonstrates how communities could revoke an SLO when they are not actively involved in development programmes and processes that affect them.

Luke (2018) provides another compelling case of how legal approvals for mining and engagement processes viewed as unfair led to the rejection of an SLO and, ultimately, the cancelled licence of a gas company, Metgasco, in the New South Wales Northern Rivers region of Australia. The first time the community of Lismore learned about the coal gas mining project was when a group of women noticed a drill rig had appeared across the valley to drill on a neighbour's property. The women became part of an anti-coal gas movement that used rallies and protests to demonstrate its opposition to the approval of Metgasco's licence to mine. Interestingly, Luke (2018) reported that the chief executive of Metgasco viewed a social licence as "an opportunity for NIMBYs [not in my backyard] to complain." In 2015, a year that became known for the Bentley blockade – thousands of people camped on the bordering property of a farmer who had signed a contract with the mining company (ibid.) – the NSW government paid Metgasco A$25 million as compensation for its cancelled gas licence (Nicolls, 2015; Luke, 2018).

Third is the pattern of repressive violence associated with mining in South Africa, including the assassination, intimidation, and physical attacks on local activists who have been vocal about the rights of the community (Hallowes & Munnik, 2016). Under such strenuous contexts, community engagement is unlikely to be useful as community members will not be free to share ideas, take initiatives, or articulate their needs. Insofar as trust is concerned, community members may resolve the question of how they should interpret the decisions of the mining company or CDT by relying on perceived procedural fairness. In most instances, they will react more negatively towards the outcomes of the decisions (and oppose an SLO) if they believe unfair tactics (i.e., intimidation and physical harm), in particular, unjust procedures, are being used.

One layer of complexity related to procedural fairness and granting an SLO that warrants further explanation is the question of who is to be considered as part of what is termed "community" or, simply, "legitimate stakeholders" (Sinclair, 2014; Gallois et al., 2017). One South African mining company executive, for example, complained about the heterogeneity of communities around mines with different needs and therefore different priorities (Benton, 2018) – a situation that prevails even among Aboriginal communities of Australia (Maddison, 2009). Others have questioned the notion of community as it relates to a social licence (Benton, 2018; Federation for a Sustainable Environment, 2018).

These views, it is contended, are a caricature of community engagement processes because, if valid, the same would apply to a formal licence to operate. Following this logic, issues of procedural fairness in community engagement processes are more often than not taken into consideration. For example, policy frameworks exist within industrialised countries (Australia and Canada, for example) and developing countries (Ghana and South Africa, for example), with specific plans for community engagement, in general, and the need to identify who should be contacted by the industry during the engagement process, in particular (DMR, 2010; Dupuy, 2017; Victoria State Government, 2017). Presumably, the inherent challenges in the identification of a broad cross-section of stakeholders – those who may be directly affected or less directly affected – apply to both the legal licence and an SLO.

Gallois et al. (2017) have made several useful observations insofar as the appeal of SLOs is concerned. They argued that an SLO speaks to "a metaphorical boundary defining the point at which development activity becomes unacceptable" (ibid. 51). Thus, legitimate stakeholder groups are geographically distributed beyond the metaphorical "fence line" of operation. Again, there is a tendency of companies to ignore external factors, including commodity prices and national agreements, that influence whether and how a mining operation conducts business (ibid.). Beyond this general concern about how mining companies could gain acceptance from local communities, elite capture is one of the ranges of practical problems confronting community development projects.

Theories of elite capture in the context of CDTs

The previous sections explained how to understand the community acceptance of mining, which involves the interplay between three factors (social impact, company contact, and procedural fairness) and trust in explaining how an SLO is obtained or maintained. Additionally, the sections identified the primary actors – mining company and CDT representatives, and community members – and their explicit and implicit agency roles in the process of seeking and consenting to a social licence. The government is a central actor in the process of obtaining an SLO, as it has the responsibility to establish an enabling (legal) environment for agreement-making between mining companies and communities. The conditions for an SLO depend primarily on the willingness of the government, through legislation and regulation, to grant or withhold access to mining companies to operate in resource-extraction projects (De Jong & Humphreys, 2016).

I propose to broaden the definitional scope of the concept of an SLO by applying insights from elite theories, in general, and elite capture at local, in particular. Such a focus can provide useful explanations on how other primary actors in the SLO process, other than those mentioned above, can influence this process. In this regard, a highly localised approach is followed by focusing only on social elites and traditional leaders. However, how does elite capture shape an SLO? That is the question addressed in the next section, by analysing the role of elites within community-based development programmes through the lens of trust.

As typified by the work of some scholars (see, for example, Thompson & Boutilier, 2011; Moffat & Zhang, 2014), the paths to an SLO are complex and time-consuming, as considerable time is needed to build trusting relationships. An elite capture analysis of an SLO, in particular, recognises the interrelatedness of various components of the community engagement processes and the general tendency of mining companies to depend on partnerships with actors within a pre-existing social structure (Rigon, 2014). Pre-existing structures within communities have, however, the potential to reproduce and reinforce relations of domination and subordination between elites and non-elites (Lund & Saito-Jensen, 2013).

Extensive theoretical discussions of the extant literature on elite theories are somewhat beyond the scope of this section. However, it is worth noting that, broadly, elites are described regarding the influence wielded, which enables them to evade public accountability (Wong, 2010). Persha and Andersson (2014: 265) define elite capture as "the process by which local elites – individuals with superior political status due to economic, educational, ethnic, or other social characteristics – take advantage of their positions to amass a disproportionately large share of resources or a flow of benefits." Demmers (2012) emphasised the dominant role of elites in dictating the terms of local decision-making processes. This notion of elite capture is similar to Musgrave and Wong's (2016: 92), which denotes how non-elites are disadvantaged by elites in the distribution of resources, project implementation, and decision-making processes. In this regard, the three categories of elites – economic, political, and social (Yamokoski & Dubrow, 2008; Musgrave & Wong, 2016) – discussed in Chapter 1, provide useful insights about the sources of power for elites.

Several scholars have contended that the fundamental shifts in political, economic, and social institutions over the past several decades led to new ways of operating which warrant a new kind of elite (Wedel, 2009; Abbink & Salverda, 2013; Davis & Williams, 2017; Wedel, 2017). Departing from the static and fixed groups in which elite studies have treated elites, Wedel (2017) advances an argument about contemporary influence elites, defined the way they operate instead. Influence elites, including traditional leaders, have adopted flexible ways of taking advantage of the renaissance of decentralised institutional decision-making arrangements in many parts of the world. Wedel (2017: 124) defines influence elites according to four distinct ways they act. First is their flexible, shifting, and overlapping roles, and a lack of a fixed attachment to any particular sector or organisation in pursuit of their strategic goals. Second is their informality and supplanting of formal structures and processes. They can use the structures and processes when they seem beneficial. Third is the myriad entities they mobilise. Fourth is their role as connectors, position in the official, corporate, private organisational ecosystem (including the above entities), and networks vis-à-vis each other (ibid.).

Elite capture and SLOs in the mining industry

Many of the ideas of the elite capture theories can be applied to the construction of an SLO between a mining company and a community. Elites, as scholars define them, function as a linkage between mining companies and communities. The strategies elites employ in gaining access to and control of resources influence how an SLO is constructed, maintained, or revoked, particularly in the context of community-based natural-resource management programmes. It is increasingly recognised, as Arnall et al. (2013) argued, that local institutional arrangements and power configurations in such localised programmes often overwhelm the customary practices of the intervening agencies. It is argued that local elite capture, together with limited transparency and accountability in these programmes, has adverse outcomes for trust in mining companies.

Research underscores the importance of community participation in local development processes but has equally noted that elite capture undermines the outcomes of these processes (Ribot, 2006; Dill, 2009; Arnall et al., 2013; Rigon, 2014, Matebesi, 2017). In a context in which patronage relationships at the local level are considered a constant feature of African politics (Platteau, 2004), limited resources to deal with power imbalances within communities are most likely to foster tendencies towards elite capture (Rigon, 2014). Elite capture occurs when project funds are controlled by an elite, often not reaching those for whom they are intended (Carrick-Hagenbarth, 2016).

There are a range of possible ways in which political elites (or government officials) can influence the conditions under which an SLO is constructed. One primary way relates to mining approval processes, which entail decisions about when, where, and under what circumstances mining can occur. More often than not, political elites use their positions to solicit personal favours from mining companies. A report by Transparency International (2017) notes that transparent and accountable mining can contribute to sustainable development. The report highlights that the government provides the regulatory frameworks, including the arrangements for integrity systems and institutions. Several cases have been documented globally about how mining companies influence both the policy and political agenda of the government of significant resource projects. In Australia, for example, allegations of soliciting, receiving, and concealing payments by politicians with close ties to the mining industry have, in some cases, led to convictions and imprisonment (Transparency International Australia, 2017). In South Africa, where state capture has become entrenched, it has been shown how public officials have taken bribes in awarding contracts to the private sector in a web of corruption involving many multinational companies (Transparency International Secretariat, 2017). Under such conditions, there will be no obligation for a mining company to adhere to the agreements concluded with a community. The focus now turns to the micro-level social dynamics of an SLO in the context of elite capture.

Traditional elites and SLOs

Local actors, including traditional leaders, have the power either to reinforce or to contradict equitable governance arrangements like, for example, CDTs. Insofar as community development funds in Africa are concerned, it has been found that their ability to uplift mining communities through improved incomes, social services, and infrastructure tends to be undermined by local power dynamics. In Ghana, one of the few countries in Africa to have developed and upheld policies for redistributing its mining wealth directly to communities (Dupuy, 2017), it was found that the traditional leaders do not often use the mineral revenue distributed to the grassroots for community development, but for personal gain instead (Standing & Hilson, 2013). In rural Mozambique, a study of non-governmental organisation intervention in community development found that non-elites are playing an essential role in monitoring and controlling leader activities. In fact, they were found to be challenging individuals in authority structures rather than the dominant structures themselves (Arnall et al., 2013).

In Indonesia (and indeed in other countries), Parani (2016) found that the efforts of kaolin mining companies in building and maintaining relations with stakeholders, for obtaining a social licence, are negotiated at meetings initiated by the village head. The study shows how community leaders enact agency in negotiating the competing interests of companies and local communities. Interestingly, as in the case of Arnall et al. (2013), the Indonesian study further demonstrates how the relationships built between the mining companies and the community leaders have become transactional (i.e., financial incentives for the leader), and serve as the determining factor.

In South Africa, the Constitution recognises customary law as a corresponding component of the legal order together with the common law (Nhlapo, 2017). Under customary law, traditional leaders have control over communal land (Pickering & Nyapisi, 2017), and this has severe consequences for the administration of benefit sharing in traditional villages, as traditional leaders have entered into complex arrangements with mining companies (Mnwana, 2015). In this regard, the Platinum Belt – an area spanning three provinces and home to 80% of the world's known platinum reserves – provides illustrative cases of how ordinary villagers have been sidelined by traditional leaders and mining companies in agreements about benefit sharing.

In the North West Province of South Africa, mining companies do not have to engage directly with rural residents before turning communal grazing and ploughing fields into massive open pits and multiple shafts. According to Mnwana (2015), the objective is to obtain the approval of the traditional leader and the tribal council. As Gallois et al. (2017) notes, where trust is lacking in a communication source (for example, a tribal authority representative), technical language from a mining company may intensify intergroup tension and contribute to conflict. Furthermore, the tendency of mining companies to demand excessive secrecy over the terms of benefit-sharing agreements that have been concluded with traditional leaders would negatively affect how the community views the companies. In some instances, when the broader community is consulted about the benefit-sharing agreements, mining companies have been blamed for negotiating in bad faith in the way in which they exploit local institutional fragilities (Wynberg & Hauck, 2014; Pickering & Nyapisi, 2017).

Thus far, the Royal Bafokeng Nation Development Trust (RBNDT) (see Chapter 4) in South Africa, which is administered by the Royal Bafokeng Traditional Authority, has demonstrated that royalties and dividends from mineral resources, if managed well, can contribute to sustainable futures for mining communities. Particularly relevant here is that the Bafokeng Nation has, over a 98-year period, been characterised by several attempts by different governments and mining companies in South Africa to dispossess it of its land rights. Unlike the opulent lifestyle based on extensive consumerism that is followed by other traditional authorities, the key of the Bafokeng model was to convert most of the royalties into minor share ownership in mining ventures, including long-term and prudent diversified investments (RBNDT, 2018). As of December 2017, RNBDT's community investment company – Royal Bafokeng Holdings – was managing a portfolio with a net asset value of approximately R32 billion (De Klerk, 2018).

Elite capture and SLOs within the context of local government participatory processes

The South African local government system is framed within a participatory evidenced planning setting as mandated by various international statutes. Over the past two decades, there has been a progressive shift towards a more sophisticated approach of public participation in local budgetary processes. While these local participatory processes form an increasingly favourite and vital part of local governance, they have not been without criticism. In particular, the process is a manifestation of political participation (Struić & Bratić, 2018) concerned with "action by citizens which is aimed at influencing decisions which are, in most cases, taken by public representatives and officials" (Parry et al., 1992: 16). In the context of South Africa, one should not look further than the high prevalence of community protests targeting local municipalities about the perceived inadequate provision of essential services, including a lack of genuine consultation over a myriad of issues (Alexander, 2010; Matebesi, 2017).

In South Africa, participatory evidenced planning takes place within the ambit of integrated development plans (IDPs), which serves as a comprehensive framework for local development over five years. The IDPs are reviewed annually and had to be aligned with the priority goals of municipalities. In this regard, the literature sheds some light on the complexities of transitioning through this process (Barichievy et al., 2005; Booysen, 2009; Piper & Nadvi, 2010; Mubungizi & Dassah, 2014; Matebesi, 2017) and, in particular, the role of elites in capturing the process. For example, as indicated before, insofar as the formal licence to operate is concerned in South Africa, mining companies must submit an SLP when applying for mining rights. This SLP is also required to be aligned to the municipal IDP, with the primary goal of an investment opportunity, economic growth, poverty reduction, and infrastructure development.

Thobatsi (2014) highlighted a few notable findings about the SLP-IDP alignment process, which, in reality, can be equally applicable to the construction of an SLO within local participatory governance structures. First, there seems to be a reluctance on the part of local councillors to be part of the IDP process. Second, the different and conflicting ideologies of political parties within local councils often cause conflicts which delay and undermine the IDP process. Partisanship is also responsible for the tendency of ward councillors to fight for the allocation of local resources to their wards as an indication of their commitment to service delivery. Third, there is a growing demand for a portion of such royalties and taxes to be paid directly to the municipalities to improve the local economic development and infrastructure challenges. Fourth, mining companies do not often know how to deal with the intersection of community expectations and demands, capacity constraints of local government and the DMR, as well as poor stakeholder engagement (ibid.).

The IDP process is a relative short process and, thus, cannot necessarily change the underlying power dynamics that allow elites to capture such processes (Lund & Saito-Jensen, 2013). Lucas's (2016) analysis of leadership, elite capture, and corruption suggested that local leaders use their positions to monopolise the planning and management of projects that were explicitly intended to incorporate participatory and accountability features. Others, he further argued, may use their control of projects to full community benefit. There is also considerable friction between villagers and elites, as well as among members of the local elite themselves over control of local resources. Furthermore, he emphasised that the "differences in the structure of these cross-cutting internal relationships and of ties between local authorities and outside government and non-government agents largely explain the differences in the degree of elite capture and its outcomes in the two cases" (Lucas, 2016: 287).

Sheeley (2016) noted that a primary form of elite capture is when political elites disregard the priorities identified by citizens in planning meetings and implement some other project. In the context of SLO construction, the actions of political leaders are a primary determinant of mining company legitimacy. If elites within the IDP engagement processes act within the interest of the community, there is an excellent possibility that the community may view the mine as a trustworthy. Conversely, if the elites disregard the input of the community relating to mining benefit sharing, or circumvent the spirit and intent of the participatory process, the potential for heightened tension between the community and mining company is real.

This problem is made possible because, often, governments and civil society organisations may only partially represent the interests of the ultimate target group for the mining benefits. In circumstances when the beneficiary, a community, does not feel that it has ownership of an intervention, its resulting lack of participation may undermine the effectiveness of such programmes. Swidler and Watkins (2009) suggested the implicit assumption in participation literature that ownership of a development activity is central to voluntary participation in its design and implementation. The tension between local government representatives and community stakeholders, including civil society organisations, one may contend, will largely depend on the nature of the mining company–community relations outside of the local participatory processes. This problem may relate to the blurred boundaries between a local government process and a mining company which endeavours to comply with the conditions of its licence or even the power of leaders within interest groups to frame issues.

Evolutionary governance theory and path dependency

The preceding sections provide insight about how an SLO is created, maintained, and lost. I now turn my focus to the EGT and path dependency. In an ever-changing global policy environment, it is vital to understand governance evolution, its processes, driving forces, and mechanisms (Van Assche et al., 2016). The EGT perspective is further necessary because it enables us to understand a dynamic network of actors, discourses, and institutions.

According to Van Assche et al. (2016: 20), governance refers to the "taking of collectively binding decisions for a community in a community, by governmental and other actors." Governance also includes various institutions, changing relationships, and governance paths. Van Assche et al. (2016: 29) state that within EGT "governance paths are histories of confrontations between these different versions of the world and different attempts to steer, govern and coordinate."

The EGT identifies three evolutionary paths: governance paths, dependencies, and path creation (Beunen et al., 2016). My interest here lies in path dependency, the value of which, for policy studies, has been argued before (Kay, 2005). There are three distinguishing levels of institutions: macro or constitutional level, policy level, and decision-making level (ibid.).

In reality, path dependency helps one understand how "legacies from the past shape future options" (Van Assche et al., 2016: 28) and explains how "successive generations of political and social actors have difficulty in departing from patterns set by their predecessors" (Crouch, 2010: 112). However, path dependency does not provide one with a means of predicting the future by analysing the past (North, 1990). In research, path dependency rather helps one to explain and compare alternatives.

The analytical understanding of path dependence can be extended to SLO studies. There are many theoretical justifications for an examination of social licensing in mining using path dependency. In a paper on path dependency in marketing systems, Layton and Duffy (2018) identified three interdependent ways of approaching path dependence: narrative, partial, and strong approaches. The first approach establishes in narrative terms what happened, the second identifies the key path dependencies in a partial analysis, and the third draws on a detailed or strong understanding of the causal dynamics at work.

A major value of path dependency within EGT helps us understand how successive generations of government policies and actions directed with a specific goal and underpinned by the hegemonic social values created a profoundly path-dependent system (Matebesi & Marais, 2018). The foundation of this path dependency resides with the processes of enforcing an SLO through CDTs in South Africa. Path dependence is seen as a natural outcome of poor decisions by people in highly complex situations who tend to neglect future developments (Koch et al., 2009; Haasnot et al., 2013). According to Parsons et al. (2019), efforts to break path dependency involve the formal recognition of the governance, values, and knowledge within policies of the subjects of study. Another solution to breaking path dependence is offered by Hämäläinen and Lahtinen (2016) who suggest different paths to resolve problem situations with multiple decision-makers and stakeholders holding different preferences and views.

Conclusion

The focus in this chapter has been on how an SLO can be constructed, maintained, and lost through paths of dependency. This chapter shows that an SLO is better understood as how to trust key variables in the context of an SLO – namely, social impacts, contact quality, and procedural fairness, as identified by Moffat and Zhang (2014). It expands these variables by including the capacity of elite capture to shape the opinion of communities towards mining companies and, thus, the conditions under which communities grant SLOs.

The literature is replete with examples of how mining companies face protests, interrupted operations, and revenue loss. The cause of such mobilisation against mining companies lies like the relationship with communities and key stakeholders. It is evident that even in cases in which an SLO has already been obtained, community engagement should be an ongoing process to create and maintain a positive and stable relationship between mining companies and communities based on trust. In South Africa, however, the trend of purging community activists who are advocates of community engagement, as exemplified in the intimidation and killing of some activists in Limpopo, is regrettable.

Similarly, CDTs under the leadership of traditional authorities and the requirement to align community-based development with the local government planning processes are among the significant challenges for mining company–community relations. The alignment process is a planning paradox for mining companies, and it reflects the failure of the mining industry to adapt to the complicated settings of the 21st-century community engagement mechanisms and structures. For example, whereas most mining companies express a desire to establish, enhance, and maintain relationships with communities, in reality they seem less concerned about investing time in the processes. Moving away from engagement efforts which are more insidious (for example, deciding philanthropic efforts in boardrooms), mining companies can ameliorate distrust among communities through the complete implementation of community development projects. Such an approach can ensure that mining companies obtain a social licence.

References

Abbink, J. & Salverda, T. (eds). (2013). *The anthropology of elites: Power, culture, and the complexities of distinction.* New York: Palgrave Macmillan.

Adali, S. (2013). *Modelling trust contexts in networks.* London: Springer.

Alexander, P. (2010). Rebellion of the poor: South Africa's service delivery protests – A preliminary analysis'. *Review of African Political Economy,* 37(123): 25–40.

Almond, G.A. & Verba, S. (1963). *Civic culture.* Princeton: Princeton University Press.

Arnall, A., Thomas, D.S.G., Twyman, C. & Liverman, D. (2013). NGOs, elite capture and community-driven development: Perspectives in rural Mozambique. *Journal of Modern African Studies,* 51(2): 305–330.

Banfield, E.C. (1958). *The moral basis of a backward society.* New York: Free Press.

Barlow, F.K., Paolini, S., Pedersen, A., Hornsey, M.J., Radke, H.R.M., Harwood, J., & Sibley, C.G. (2012). The contact caveat: Negative contact predicts increased prejudice more than positive contact predicts reduced prejudice. *Personality and Social Psychology Bulletin*, 38: 1629–1643.

Barichievy, K., Piper, L. & Parker, B. (2005). Assessing 'participatory governance' in local government: A case-study of two South African cities. *Politeia*, 24(3): 370–393.

Bauer, P.C. & Freitag, M. (2018). Measuring trust. In: E.M. Uslaner (ed.). *The Oxford handbook of social and political trust* (pp. 15–36). New York: Oxford University Press.

Benton, N. (2018). What will it take to trust the mining industry?. Available at https://medium.com/@n63n70n/what-will-it-take-to-trust-the-mining-industry-d7633aba6eca (accessed 14 June 2019).

Bice, S. (2014). What gives you a social licence? An exploration of the social licence to operate in the Australian mining industry. *Resources*, 3(1): 62–80.

Black, L. (2013). *The social licence to operate: Your management framework for complex times.* Oxford: Do Sustainability.

Booysen, S. (2009). Beyond the ballot and the brick: Continuous dual repertoires in the politics of attaining service delivery in South Africa? In: A. McLennan & B. Munslow (eds). *The politics of service delivery* (pp. 104–136). Johannesburg: Wits University Press.

Bond, P. & Mottiar, S. (2013). Movements, protests and a massacre in South Africa. *Journal of Contemporary African Studies*, 31: 283–302.

Boutilier, R. & Thompson, I. (2011). *Modelling and measuring the social licence to operate: Fruits of a dialogue between theory and practice* [online]. Available at http://socialicense.com/publications/Modelling%20and%20Measuring%20the%20SLO.pdf (accessed 14 January 2017).

Boutilier, R.G., Black, L. & Thompson, I. (2012). *From metaphor to management tool: How the social license to operate can stabilise the socio-political environment for business.* Carlton, Victoria, Australia, 20 November. Australian Institute of Mining and Metallurgy, pp. 227–237.

Brueckner, M., Durey, A., Mayes, R. & Pforr, C. (2014). Living the resource boom. In: M. Brueckner, A. Durey, R. Mayes & C. Pforr (eds). *Resource curse or cure? On the sustainability of development in Western Australia* (pp. 221–222). Heidelberg: Springer Verlag.

Carrick-Hagenbarth, J. (2016). Unpublished Doctoral Thesis. Amherst: University of Massachusetts [online]. Available at https://scholarworks.umass.edu/cgi/viewcontent.cgi?article=1793&context...2 (accessed 30 June 2018).

Carstens, J. & Hilson, G. (2009). Mining, grievance and conflict in rural Tanzania. *International Development Planning Review*, 31(3): 301–326.

Cohen, M.A. (2016). The question of public trust in business. Comments on Harris, J.D., Moriarty, B.T. & Wicks, A.C (eds). (2014). *Public Trust in Business*. Cambridge: Cambridge University Press. *Journal of Trust Research*, 6(1): 96–103.

Cohen, M.A. & Dienhart, J.J. (2013). Moral and amoral conceptions of trust, with an application in organizational ethics. *Journal of Business Ethics*, 112(1): 1–13.

Cole, A., Fox, S., Pasquier, R. & Stafford, I. (2018). Political trust in France's multi-level government. *Journal of Trust Research*, 8(1): 45–67.

Costanza, J.N. (2016). Mining conflict and the politics of obtaining a Social License: Insight from Guatemala. *World Development*, 79(C): 97–113.

Christ, O., Schmid, K., Lolliot, S., Swart, H., Stolle, D., Tausch, N. et al. (2014). Contextual effect of positive intergroup contact on outgroup prejudice. *PNAS*, 111(11): 3996–4000.

Chuhan-Pole, P., Dabalen, A.L. & Land, B.C. (2017). *Mining in Africa: Are local communities better off?* Washington, DC: World Bank.

Dagvadorj, L., Byamba, B. & Ishikawa, M. (2018). Effect of local community's environmental perception on trust in a mining company: A case study in Mongolia. *Sustainability*, 10: 614.

Daily Maverick. (2016). amaBhungane: Broken Trust in Mapela – The people, the kgoshi and the cool R175m. *Daily Maverick*, 10 May [online]. Available at www.dailymaverick. co.za/article/2016-05-10-amabhungane-broken-trust-in-mapela-the-people-the-kgoshi-and-the-cool-r175m/# (accessed 14 January 2017).

Dalton, R.J. (2005). The social transformation of trust in government. *International Review of Sociology*, 15(1): 129–145.

Davis, A. & Williams, K. (2017). Introduction: Elites and power after financialization. *Theory, Culture & Society*, 34(5–6): 3–26.

De Jong, W. & Humphreys, D. (2016). A failed social licence to operate for the neoliberal modernization of Amazonian resource use: The underlying causes of the Bagua tragedy of Peru. *Forestry*, 89: 552–564.

Demmers, J. (2012). *Theories of violent conflict: An introduction*. London: Routledge.

De Klerk, F. (2018). Royal Bafokeng Mine [online]. Available at https://fransdeklerk.com/royal-bafokeng-mine/ (accessed 12 December 2018).

DMR (Department of Mineral Resources) (2010). *Guidelines for the submission of a Social and Labour Plan*. Pretoria: DMR.

Dill, B. (2009). The paradoxes of community-based participation in Dar es Salaam. *Development and Change*, 40(4): 717–743.

Dougherty, M.L. & Olsen, T.D. (2014). They have good devices: Trust, mining, and the microsociology of environmental decision-making. *Journal of Cleaner Production*, 84(1): 183–192.

Duncan, J. (2014). South African journalism and the Marikana massacre: A case study of an editorial failure. *The Political Economy of Communication*, [S.l.], 1(2) [online]. Available at www.polecom.org/index.php/polecom/article/view/22/198 (accessed 11 November 2017).

Dupuy, K. (2017). Corruption and elite capture of mining community development funds in Ghana and Sierra Leone. In: D. Williams & P. Le Billon (eds). *Corruption, natural resources and development: From resource curse to political ecology* (pp. 69–79). Cheltenham and Northampton (MA): Edward Elgar Publishing.

Economix Research (2016). The impact of platinum Mining in Rustenburg: A high-level analysis, 14 March 2016 [online]. Available at https://eunomix.com/cmsAdmin/uploads/eunomix-research_-the-impact-of-platinum-mining-in-rustenburg_march2016_001.pdf (accessed 11 November 2017).

Evans, A.M. & Krueger, J.I. (2009). The psychology (and economics) of trust. *Personality Psychology Compass*, 3(6): 1013–1017.

Federation for a Sustainable Environment (2018). *The impact of mining on the South African economy and living standards* [online]. Available at www.fse.org.za/index.php/item/593-the-impact-of-mining-on-the-south-african-economy-and-living-standards (accessed 13 October 2018).

Ferragina, E. (2009). The never-ending debate about the moral basis of a backward society: Banfield and "amoral familism." *Journal of the Anthropological Society of Oxford-online*, 1(2): 1–160 [online]. Available at https://halshs.archives-ouvertes.fr/halshs-01314030 (accessed 12 August 2018).

Fitzgerald, J. & Wolak, J. (2016). The roots of trust in local government in Western Europe. *International Political Science Review*, 37(1):130–146.

Franks, D.M. & Cohen, T. (2012). Social licence in design: Constructive technology assessment within a mineral research and development institution. *Technological Forecasting and Social Change*, 79(7): 1229–1240.

Fukuyama, F. (1995). *Trust: The social virtues and the creation of prosperity*. New York: Free Press.

Gallois, C., Ashworth, P., Leach, P. & Moffat, K. (2017). The language of science and social licence to operate. *Journal of Language and Social Psychology*, 36(1): 45–60.

Gehman, J., Lefsrud, L.M. & Fast, S. (2017). Social license to operate: Legitimacy by another name? *Canadian Public Administration*, 60(2): 293–317.

Granovetter, M. (1985). Economic action and social structure: The problem of embeddedness. *American Journal of Sociology*, 91(3): 481–451.

Haasnoot, J., Kwakke, J.H., Walker, W.E. & Maat, J. (2013). Dynamic adaptive policy pathways: A method for crafting robust decisions for a deeply uncertain world. *Global Environmental Change*, 23(2): 485–498.

Hajkowicz, S.A., Heyenga, S. & Moffat, K. (2011). The relationship between mining and socio-economic well-being in Australia's regions. *Resources Policy*, 36(1): 30–38.

Hallowes, D. & Munnik, V. (2016). *The groundWork Report 2016: Destruction of the Highveld*. Pietermaritzburg: groundWork [online]. Available at www.groundwork.org.za/reports/gWReport_2016.pdf (accessed 14 January 2018).

Hardin, R. (1992). The street-level epistemology of trust. *Analyse & Kritik*, 14: 152–176.

Harvey, S.J. Jr. (2014). You can have your trust and calculativeness, too: Uncertainty, trustworthiness, and the Williamson thesis. *Journal of Trust Research*, 4(1): 57–65.

Hämäläinen, R.J. & Lahtinen, T.J. (2016). Path dependence in operational research – How the modeling process can influence the results. 3: 14–20.

Hathaway, O.A. (2003). Path dependence in the law: The course and pattern of legal change in a common law system. *John M. Olin Center for Studies in Law, Economics, and Public Policy Working Papers*. Paper 270 [online]. Available http://digitalcommons.law.yale.edu/lepp_papers/270 (accessed 14 June 2017).

Hayward, L.E., Tropp, L.R., Hornsey, M.J. & Barlow, K.B. (2017). Toward a comprehensive understanding of intergroup contact: Descriptions and mediators of positive and negative contact among majority and minority groups. *Personality and Social Psychology Bulletin*, 43(3): 347–364.

Herain, M.N., Hamm, J.A., Tomkins, A.J. & Pytlik Zillig, L.M. (2012). Public Participation, procedural fairness, and evaluations of local governance: The moderating role of uncertainty. *Journal of Public Administration Research and Theory*, 22(4):815–840.

Hewstone, M. (2009). Living apart, living together? The role of intergroup contact in social integration. *Proceedings of the British Academy*, 162: 243–300.

Hunt, J. (2013). Engaging with Indigenous Australia – Exploring the conditions for effective relationships with Aboriginal and Torres Strait Islander communities. *Closing the Gap Clearinghouse*, Issues paper no. 5 [online]. Available at www.aihw.gov.au/getmedia/7d54eac8-4c95-4de1-91bb-0d6b1cf348e2/ctgc-ip05.pdf.aspx?inline=true (accessed 16 June 2017).

Igarashi, T., Kashima, T., Kashima, E.S., Farsides, T., Kim, U., Strack, F., Werth, L. & Yuki, M. (2008). Culture, trust, and social networks. *Asian Journal of Social Psychology*, 11(2): 88–101.

Jinnah, Z. (2017). Silence and invisibility: Exploring labour strategies of Zimbabwean farmworkers in Musina, South Africa. *South African Review of Sociology*, 48(3): 46–63.

Joyce, S. & Thompson, I. (2000). Earning a social licence to operate: Social acceptability and resource development in Latin America. *Canadian Mining and Metallurgical Bulletin*, 93(1037): 49–52.

Kauff, M., Asbrock, F., Wagner, U., Pettigrew, T.F., Hewstone, M., Schäfer, S.J. & Christ, O. (2017). (Bad) feelings about meeting them? Episodic and chronic intergroup emotions associated with positive and negative intergroup contact as predictors of intergroup behavior. *Frontiers of Psychology*, 8: 1449.

Kay, A. (2005). A critique of the use of path dependency in policy studies. *Public Administration*, 83(3): 553–571.

Kikuchi, M., Watanabe, Y. & Yamagish, T. (1997). Judgment accuracy of other's trustworthiness and general trust: An experimental study. *Japanese Journal of Experimental Social Psychology*, 37(1): 23–36.

Koch, J., Eisend, M. & Petermann, A. (2009). Path dependence in decision-making processes: Exploring the impact of complexity under increasing returns. *BuR Business Research Journal*, 2(1): 67–84.

Koivurova, T., Buanes, A., Riabova, L., Didyk, V., Ejdemo, T., Poelzer, G., Taavo, P. & Lesser, P. (2015). "Social license to operate": A relevant term in Northern European mining? *Polar Geography*, 38(3): 194–227.

Lacey, J., Carr-Cornish, S., Zhang, A., Eglinton, K. & Moffat, K. (2017). The art and science of community relations: Procedural fairness at Newmont's Waihi Gold operations, New Zealand. *Resources Policy*, 52: 245–254.

Layton, R. & Duffy, S. (2018). Path dependency in marketing systems: Where history matters and the future casts a shadow. 38(4).

Lelyveld, J. (1981). At times South Africa's mines, race are rigid. *The New York Times* [online]. Available at www.nytimes.com/1981/06/09/world/at-south-africa-s-mines-race-barriers-are-rigid.html.

Levi, M. & Stoker, L. (2000). Political trust and trustworthiness. *Annual Review of Political Science*, 3(1): 475–507.

Lucas, A. (2016). Elite capture and corruption in two villages in Bengkulu Province, Sumatra. *Human Ecology: An Interdisciplinary Journal*, 44: 287–300.

Luke, A. (2018). Not getting a social licence to operate can be a costly mistake, as coal seam gas firms have found. *The Conversation* [online]. Available at https://theconversation.com/not-getting-a-social-licence-to-operate-can-be-a-costly-mistake-as-coal-seam-gas-firms-have-found-93718 (accessed 16 March 2019).

Lund, J.F. & Saito-Jensen, M. (2013). Revisiting the issue of elite capture of participatory initiatives. *World Development*, 46: 104–112

Maccannon, B.C., Asaad, C.T. & Wilson, M. (2017). Contracts and trust: Complements or substitutes. *Journal of Institutional Economics*, 14(5): 811–832.

Maddison, S. (2009). *Black politics: Inside the complexity of Aboriginal political culture*. Crows Nest: Allen & Unwin.

Maluleke, G. & Pretorius, L. (2016). Modelling the impact of mining on socio-economic infrastructure development – A system dynamics approach. *South African Journal of Industrial Engineering*.

Mancini, L. & Sala, S. (2018). Social impact assessment in the mining sector: Review and comparison of indicators frameworks. *Resources Policy*, 57: 98–111.

Matebesi, S.Z. (2017). *Civil strife against local governance: Dynamics of community protests in contemporary South Africa*. London: Barbara Budrich.

Mayes, R. (2014). Mining and (sustainable) local communities: Transforming Ravensthorpe, Western Australia. In: M. Brueckner, A. Durey, R. Mayes & C. Pforr (eds). *Resource curse or cure? On the sustainability of development in Western Australia* (pp. 223–237). Heidelberg: Springer-Verlag.

McKeown, S. & Psaltis, C. (2017). Intergroup contact and the mediating role of intergroup trust on outgroup evaluation and future contact intentions in Cyprus and Northern Ireland. *Peace and Conflict: Journal of Peace Psychology*.

Meesters, M.E. & Behagel, J.H. (2017). The social licence to operate: Ambiguities and the neutralization of harm in Mongolia. *Resources Policy*, 53(September): 274–282.

Merten, M. (2016). MPs in budget vote conveyer belt. *Daily Maverick*, 19 April [online]. Available at www.dailymaverick.co.za/article/2016-04-19-parliamentary-diary-mps-in-budget-vote-conveyer-belt/#.W0q-8bh9iUk (accessed 17 July 2017).

Mnwana, S. (2015). Mining and "community" struggles on the Platinum Belt: A case of Sefikile village in the North West Province, South Africa. *The Extractive Industries and Society*, 2(3): 500–508.

Moffat, K. & Zhang, A. (2014). The paths to social licence to operate: An integrative model explaining community acceptance of mining. *Resources Policy*, 39(March): 61–70.

Mubungizi, B.C. & Dassah, M. (2014). Public participation in South Africa: Is intervention by the courts the answer? *Journal of Social Science*, 39(3): 275–284.

Musgrave, M.K. & Wong, S. (2016). Towards a more nuanced theory of elite capture in development projects. The importance of context and theories of power. *Journal of Sustainable Development*, 99(3): 87–103.

Newton, K., Stolle, D. & Zmerle, S. (2018). Social and political trust. In: E.M. Uslaner (ed.). *The Oxford handbook of social and political trust* (pp. 37–56). New York: Oxford University Press.

Nhlapo, T. (2017). Customary law in post-apartheid South Africa: Constitutional confrontations in culture, gender and "living law". *South African Journal on Human Rights*, 33(1): 1–24,

Nicolls, S. (2015). NSW government set to pay $25 million for Metgasco north coast gas licences [online]. Available at www.smh.com.au/national/nsw/nsw-government-set-to-pay-25-million-for-metgasco-north-coast-gas-licences-20151102-gkodbr.html (accessed 28 May 2017).

North, D.C. (1990). *Institutions, institutional change and economic performance*. Cambridge: Cambridge University Press.

O'Faircheallaigh, C. (2011). Indigenous women and mining agreement negotiations: Australia and Canada. In: K. Lahiri-Dutt (ed.). *Gendering the field: Towards sustainable livelihoods for mining communities* (pp. 87–110). Canberra: ANU E Press.

Opp, K. (2009). *Theories of political protest and social movements: A multidisciplinary introduction, critique, and synthesis*. New York: Routledge.

Owen, J. & Kemp, D. (2013). Social licence and mining: A critical perspective. *Resources Policy*, 38(1): 29–35.

Owen, J. & Kemp, D. (2014). Mining and community relations: Mapping the internal dimensions of practice. *Extractive Industries and Society*, 1(1): 12–19.

Owen, J. & Kemp, D. (2017). *Extractive relations: Countervailing power and the global mining industry*. New York: Routledge.

Parani, R. (2016). Community leaders and community relations practitioners as agents for corporate interests: A case study of Indonesian mining. Doctoral thesis, RMIT University. Melbourne: RMIT University.

Parker, G., Parker, R. & Towner, T.L. (2014). Rethinking the meaning and measurement of political trust. In: C. Eder, I.C. Mochmann & M. Quandt (eds). *Political trust and disenchantment with politics: International perspectives* (pp. 59–82). Leiden: Brill.

Parry, G., Moyser, G. & Day, N. (1992). *Political participation in Britain*. Cambridge, MA: Cambridge University Press.

Parsons, R. & Moffat, K. (2014). Constructing the meaning of social licence. *Social Epistemology*, 28(3–4): 340–363.

Parsons, M., Nalau, J., Fisher, K. & Brown, C. (2019). Disrupting path dependency: Making room for Indigenous knowledge in river management. *Global Environmental Change*, 56: 95–113.

Persha, L. & Andersson, K. (2014). Elite capture risk and mitigation in decentralized forest governance regimes. *Global Environmental Change*, 24: 265–276.

Pettigrew, T.F. & Hewstone, M. (2017). The single factor fallacy: Implications of missing critical variables from an analysis of intergroup contact theory. *Social Issues Policy Review*, 11: 8–37.

Pettigrew T.F. & Tropp, L.R. (2011). *When groups meet: The dynamics of intergroup contact.* Philadelphia, PA: Psychology Press.

Pickering, J. & Nyapisi, T. (2017). A community left in the dark: The case of Mapela. Struggles for transparency and accountability in South Africa's Platinum Belt. In: *Good Company* (2nd Edition, pp. 27–35). Conversations around transparency and accountability in South Africa's extractive sector by Open Society Foundation for South Africa. Cape Town: OSF.

Piper, L. & Nadvi, L. (2010). Popular mobilisation, party dominance and participatory governance in South Africa. In: L. Thompson & C. Tapscott (eds). *Citizenship and social movements: Perspectives from the global South* (pp. 212–238). London: Zed Books.

Platteau, J.P. (2004). Monitoring elite capture in community-driven development. *Development and Change*, 35(2): 223–46.

Poppo, L. & Schepker, D.J. (2010). Repairing public trust in organizations. *Corporate Reputation Review*, 13(2): 124–141.

Prno, J. & Scott Slocombe, D. (2012). Exploring the origins of "social license to operate" in the mining sector: Perspectives from governance and sustainability theories. *Resources Policy*, 37(3): 346–357.

Putnam, R. (2000). *Bowling alone: The collapse and revival of American community.* New York, NY: Free Press.

Reimer, N.K., Becker, J.C., Benz, A., Christ, O., Dhont, K., Klocke, U., et al. (2017). Intergroup contact and social change: Implications of negative and positive contact for collective action in advantaged and disadvantaged groups. *Personality and Social Psychology Bulletin*, 43(1): 121–136.

Ribot, J.C. (2006). Choose democracy: Environmentalists socio-political responsibility. *Global Environmental Change*, 16(2): 115–119.

Rigon, A. (2014). Building local governance: Participation and elite capture in slum-upgrading in Kenya. *Development and Change*, 45(2): 257–283.

Ripperger, T. (1998). *Ökonomik des Vertauens.* Tübingen.

Rosenberg, M. (1956). *Misanthropy* and political ideology. *American Sociological Review*, 21(6): 690–695.

Rousseau, D.M., Sitkin, S.B., Burt, R.S. & Camerer, C. (1998). Not so different after all: A cross discipline view of trust. *The Academy of Management Review*, 23(3): 393–404.

RBNDT (Royal Bafokeng Nation Development Trust) (2018). RBNDT [online]. Available at www.bafokeng.com/organisation/entities/rbndt (accessed 30 February 2019).

Sarkar, S., Gow-Smith, A, Morakinyo, T., Frau, R. & Kuniholm, M. (2010). *Mining community development agreements – Practical experiences and field studies.* Washington: World Bank.

Sheeley, R. (2015). Mobilization, participatory planning institutions, and elite capture: Evidence from a field experiment in rural Kenya. *World Development*, 67: 251–266.

Standing, A. & Hilson, G. (2013). Distributing mining wealth to communities in Ghana: Addressing problems of elite capture and political corruption. Anti-corruption Resource Centre (U4 Issue 5) [online]. Available at www.cmi.no/publications/file/4791-distributing-mining-wealth-to-communities-in-ghana.pdf (accessed 29 March 2018).

Steeves, H.L. & Melkote, S.R. (2015). *Communication for development: Theory and practice for empowerment*. New Delhi: Sage.

Stokes, D. (1962). Popular evaluations of government: An empirical assessment. In H. Cleveland & H.D. Lasswel (eds). *Ethics and bigness: Scientific, academic, religious, political and military* (pp. 61–72). New York: Harper.

Struić, G. & Bratić, V. (2018). Public participation in the budgetary process in the Republic of Croatia. *Public Sector Economics*, 42(1): 67–92.

Swidler, A. & Watkins, S.C. (2009). "Teach a man to fish": The sustainability doctrine and its social consequences. *World Development*, 37(7): 1182–1196.

Tekwa, E.W., Fenichel, E.P., Levin, S.A. & Pinsky, M.L. (2019). Path-dependent institutions drive alternative stable states in conservation. *PNAS*, 116(2): 689–694.

Terwindt, C. & Schliemann, C. (2017). Tricky business: Space for civil society in natural resource struggles. Publication Series on Democracy. Berlin: Heinrich Böll Foundation or the European Center for Constitutional and Human Rights [online]. Available at www.boell.de/sites/default/files/tricky-business.pdf (accessed 15 March 2018).

Thobatsi, J. (2014). *The alignment of Social and Labour Plan (SLP) commitments with municipal Integrated Development Plans (IDPs)*. Unpublished mini-dissertation, Potchefstroom Campus of the North-West University. Potchefstroom: North-West University [online]. Available at http://dspace.nwu.ac.za/bitstream/handle/10394/12049/Thobatsi_JT.pdf?sequence=1 (accessed 4 June 2019).

Thompson, I., Tapscott, C. & De Wet, P.T. (2018). An exploration of the concept of community and its impact on participatory governance policy and service delivery in poor areas of Cape Town, South Africa. *Politikon*, 45(2): 276–290.

Thompson, I. & Boutilier, R.G. (2011). The Social License to Operate. In P. Darling (ed.). *SME mining engineering handbook* (3rd edn, pp. 1779–1790). Littleton: Society for Mining, Metallurgy, and Exploration.

Tilly, C. (2004). *Trust and rule*. New York: Columbia University Press.

Transparency International (2017). Combatting corruption in mining approvals: Assessing the risks in 18 resource-rich countries. Transparency International [online]. Available at www.transparency.org/whatwedo/publication/combatting_corruption_in_mining_approvals (accessed 16 May 2018).

Transparency International Australia (2017). Corruption risks in mining approvals: Australian snapshot. Transparency international Australia [online]. Available at https://transparency.org.au/tia/wp/2017/09/M4SD-Australia-Report_Final_Web.pdf (accessed 26 April 2018).

Transparency International Secretariat (2017). *South African MPs should vote President Zuma out of office in no confidence test* [online]. Available at www.transparency.org/news/pressrelease/south_african_mps_should_vote_president_zuma_out_of_office_in_no_confidence (accessed 12 May 2018).

Tyler, T.R. (2006). *Why people obey the law*. Princeton, NJ: Princeton University Press.

Uslaner, E.M. (2018). The study of trust. In: E.M. Uslaner (ed.). *The Oxford handbook of social and political trust* (pp. 3–14). New York: Oxford University Press.

Wang, L., Awuah-Offei, K., Que, S. & Yang, W. (2016). Eliciting drivers of community perceptions of mining projects through effective community engagement. *Sustainability*, 8(7): 1–17.

Wedel, J.R. (2009). *Shadow elite: How the world's new power brokers undermine democracy, government, and the free market*. New York: Basic Books.

Wedel, J.R. (2017). From power elites to influence elites: Resetting elite studies for the 21st Century. *Theory, Culture & Society*, 34(5–6): 153–178.

Wesley, A. & MacCallum, D. (2014). The political economy of corporate social responsibility in the resources sector in Western Australia: A case study of the proposed James Price Point LNG precinct. In: M. Brueckner, A. Durey, R. Mayes & C. Pforr (eds). *Resource curse or cure? On the sustainability of development in Western Australia* (pp. 59–73). Heidelberg: Springer Verlag.

Williams, R. & Walton, A. (2013). The Social License to Operate and coal seam gas development. A literature review report to the Gas Industry Social and Environmental Research Alliance (GISERA). CSIRO, Canberra [online]. Available at www.gisera.org.au/publications/-tech_reports_papers//Socioeco-proj_5-lit-review.pdf (accessed 29 March 2017).

Wong, S. (2010). Elite capture or capture elites? Lessons from the "counter-elite" and "co-opt-elite" approaches in Bangladesh and Ghana. *WIDER Working Paper* 2010/082. Helsinki: UNU-WIDER.

Wynberg, R. & Hauck, M. (2014). People, power, and the coast: A conceptual framework for understanding and implementing benefit sharing. *Ecology and Society*, 19(1): 27.

Yamokoski, A. & Dubrow, J.K. (2008). How do elites define influence? Personality and respect as sources of social power. *Sociological Focus*, 41(4): 319–336.

Zanini, M.T.F. & Migueles, C.P. (2013). Trust as an element of informal coordination and its relationship with organizational performance. *EconimiA*, 14(2): 77–87.

Zhang, A., Moffat, K., Lacey, J., Wang, J., González, R., Uribe, K., Cui, L. & Dai, Y. (2015). Understanding the social licence to operate of mining at the national scale: A comparative study of Australia, China and Chile. *Journal of Cleaner Production*, 108(Part A): 1063–1072.

3 Mining regulatory frameworks and civil society mobilisation

> Firstly, I need to declare that I am not an academic. Nor have I studied any of the social sciences or human behavioural fields such as psychology. Nor have I gone down the MBA route. I am a dyed-in-the-wool mining engineer with a master's degree in engineering science. However, what I do have, and what I want to share, is over 30 years' experience working with, managing and leading people, which is often a far more variable and volatile commodity than the mineral that I was trying to mine!
>
> (Gill, 2017: 60)

Introduction

The premise of the above quote is an acknowledgement of a well-known – but often ignored – key aspect of a successful enterprise: the human element. Similarly, insofar as democratic accountability is concerned, governments that are accountable to their citizens are more likely to respond to their demands than those that are not (Bjuremalm et al., 2014). On one level, this description of democratic accountability suggests that, in the context of natural-resource governance, governments will advance and safeguard the right of its citizens to involvement in decision-making processes. On another level, it points to a peculiar configuration government authority in that governments may respond and not necessarily provide genuine spaces for public participation in mining-development decision-making processes. Notwithstanding the progressive nature of South Africa's Constitution (Friedman, 2006), thus far, the post-apartheid government – as others elsewhere – has abandoned the path of collaborative relationships with civil society in favour of a neo-extractive-centric natural-resource governance approach (Terwindt & Schliemann, 2017).

The current extractive governance approach in South Africa is characterised by bilateral dialogue between the government and the mining industry. This approach not only tends to reinforce the erstwhile approach of the apartheid government to close down spaces for public participation in economic and political arenas but has – implicitly or explicitly – shifted responsibility to advance the aspirations of people to civil society. As elsewhere in the world, civil society mobilisation calling for the inclusion of mining communities and directly affected stakeholders in natural-resource governance has gained traction in the literature (Hayman, 2014; Gufstafsson, 2018). Today, South African mining operations have become airy spaces of local protests by marginalised communities, whereas a united front of civil society organisations (CSOs) wages campaigns at the national level, including challenging the government directly.

This chapter looks at the evolution of the natural-resource governance approach in South Africa, in particular how it has created new opportunities and spaces for the articulation of civil society demands for inclusion in mining-development decision-making processes. In so doing, the focus is on some crucial mining regulatory regimens, and the community development trusts (CDTs) initiated voluntarily by some mining companies. Furthermore, the chapter makes four interrelated arguments. First, it advances the narrative that due to the political decay in South Africa, punctuated with rampant corruption in various spheres of governance, the government has adopted a tokenistic approach with civil society in mining-development decision-making processes. It does this by arguing that the pressure to collect revenues in the form of mining profit-related taxes resulted in a neo-extractive development approach.

Second, it argues that, paradoxically, mining companies are quick to highlight to government the protection international investment treaties provide against the impact of resource regulation on their assets, yet they fail to obey international instruments enshrining public participation in the decision-making process. Third, it argues that the space CSOs occupy today in South Africa illustrates efforts of previously fragmented groups that demanded to participate directly in mining-development decision-making processes. Two central features of these CSOs are, first, their coalitions with international organisations and, second, they possess the type of autonomy that democracy requires. The fourth argument is that while CDTs illuminate how local-level agreements (LLAs) and benefit shares can generate some spaces for innovative community involvement in mining development, these trusts serve as conduits for mining company–community conflict.

Mining policy and practices in South Africa before 1994

Mining in South Africa was once the main driving force behind the economic power of the country relative to other African countries. The Department of Energy and the Department of Mineral Resources have the responsibility of regulating the petroleum industry and mining industry, respectively. By way of background, as in the case of Australia, for example, where the issue of aboriginal rights has been dominating (Attwood, 2003; Broome, 2010; Short, 2016), the Native Land Act 27 of 1913 – an early legislation in respect of mineral rights and mineral development in South Africa – considered the rights of indigenous communities as unimportant (Cawood & Minnitt, 1998; Lang, 1999). These authors further provide a brief discussion about how the South African Development Act 18 of 1936 was enacted in an attempt to restore the rights of indigenous

Table 3.1 Overview of pre-1994 mining legislation in South Africa

Year	Act	Purpose
1913	The Native Land Act 27	Considered the rights of indigenous communities as unimportant
1925	Reserved Minerals Development Act of 1925	A provision whereby landowners of Alienated State Land or their nominees acquired the exclusive right to prospect and mine on their land. However, if a mine were established on this type of land, the State would be entitled to royalty payments
1936	South African Development Act 18 of 1936	Creation of separate development attempts made to restore the rights of indigenous communities through the creation of tribal lands
1942	Base Minerals Development Act 39 of 1942 and its Amendment Act 22 of 1955	Gave the State the right to intervene should the owner of any land, including private land, not exercise his or her exclusive right to prospect for and mine base minerals
	Natural Oil Act 46 of 1942	Introduction of a provision whereby the State reserved the right to prospect and mine for oil for itself
1964	Precious Stones Act 73 of 1964	Vested in the State the right to mine and dispose of precious stones
1967	Mining Titles Registration Act 16 of 1967	"A tactical move on the part of government to collect information about the different categories of mineral rights ownership" (Cawood & Minnitt, 1998: 371)
	Mining Rights Act 20 of 1967	An attempt to consolidate the plethora of laws under a single Act
	Atomic Energy Act 90 of 1967	Provides that, unless expressly exempted, no person may produce or otherwise acquire or dispose of or import into or export from South Africa or have or use or convey or cause to be conveyed, any radioactive isotope, without written authorisation of the Atomic Energy Board
1975	Mineral Laws Supplementary Act 10 of 1975	Provided, first, a means by which mining companies could obtain mineral rights over land it prevented further fragmentation of private mineral rights by testamentary succession without State approval
1977	Petroleum Products Act 20 of 1977	System for allocation of licences for liquefied petroleum gas and paraffin
1991	Minerals Act 50 of 1991	The "right-to-mine" principle which had been a legal cornerstone of mineral exploitation in South Africa, expired Section 64, which allows for the outright disposal of State-owned mineral rights to the private sector. The aim was to reduce government involvement and to create a market for State-owned mineral rights

Source: Adapted from Cawood and Minnitt (1998) and Van der Schyff (2012).

communities through the creation of tribal lands, commonly known as the homeland system. Almost six decades later, the Restitution of Land Rights Act 22 of 1994 represented a drastic attempt by the post-apartheid government to offer a solution to people who had lost their land as a result of racially discriminatory practices, such as forced removals (Lang, 1999).

After the establishment of the Republic of South Africa in 1994, several notable observations can be made with regard to the complexity of divergent and convoluted regulatory regimes, in particular, the system of administering mineral rights. As Table 3.1 illustrates, first, it was the attempt to consolidate the various statutes that regulated minerals. These acts were the Precious Stones Act 73 of 1964, the Mining Rights Act 20 of 1967, the Mining Titles Registration Act 16 of 1967, and the Atomic Energy Act 90 of 1967 (Cawood & Minnitt, 1998; Van der Schyff, 2012).

Second, the primary purpose of the consolidation of the enacted legislation was to restructure the extractive industry in order to maximise revenue for the country (Van der Schyff, 2012). Third, before 1991, the reservation of certain rights relating to minerals in favour of the State was often encountered in legislation. Sometimes this reservation applied only to precious stones and metals (Van der Schyff, 2012). However, the approach of the regulation of the South African oil industry, in particular, relied more upon government-initiated agreements intended to resolve market problems than upon legislation. As a result, the downstream petroleum industry, which is regulated by the Petroleum Products Act, 1977 (Act No. 120), before its amendments in 2003 and 2005, was shrouded in secrecy. This made it difficult for those outside of the petroleum mining industry or in the South African government to acquire information, due to attempts by the government to circumvent the restrictions of sanctions against the country (Kapdi et al., 2015).

The complex system of mineral rights ownership enabled the government to reward and protect the interest of private enterprise in the exploitation of minerals in South Africa (Cawood & Minnitt, 1998). The Mineral Rights Act 50 of 1991 reorganised the mineral law of South Africa and became the legal basis for all mineral and prospecting rights in existence at that time. An essential feature of this Act, Ramatji (2013) contended, was that it amounted to a system of universal law rights. Expanding on this, Dale (1997) stated that although the holding of mineral rights by the government and private holders was continued, provision was provided for the government to alienate its mineral rights. In fact, this entrenched policy of reservation to the government of the right to mine was abolished in favour of vesting the right to prospect and mine in the holder of the mineral rights, subject to acquisition by such holder of a permit or licence authorising the holder to exercise the rights thus held.

> Not only thereby did the State relinquish the control of the right to mine but it also relinquished the right to receive a lease consideration in respect of previously granted mining leases and no provision for similar lease consideration was made in regard to the future, save where the State itself is the mineral right holder.
>
> (Dale, 1997: 16)

The primary aim of the Minerals Act of 1991 was to reduce government involvement and to create a market for State-owned mineral rights (Ramatji, 2013) and regarded as unique in the global context. Despite the fact that many racially discriminatory laws had been abolished by 1991, the Minerals Act failed to address the injustices and imbalances which had been created by the apartheid government: the move towards the privatisation of mineral resources led to increased monopolisation of the mineral wealth by predominantly white mining companies, while black communities remained sources of labour in mining areas, without seeing tangible, long-term benefits (Mitchell et al., 2012).

What are the implications of the acts above? From a civil society perspective, and black South African perspective, in particular, the issue of land ownership played a significant role in the approach of various governments before 1994. Central in depriving black South Africans of the rights to own and administer their own affairs were statutes such as the Black Land Act 27 of 1913 and the Development Land and Trust Act 18 of 1936 (Miller & Pope, 2000; Hall, 2010; Mitchell et al., 2012). It is thus not surprising that the country's land reform programme, which aims to redress the racial imbalance in landholding and secure the property rights of historically disadvantaged people (Department of Land Affairs, 1997; Lahiff, 2007), is currently at the forefront of the plans to redistribute land without compensation outlined by President Cyril Ramaphosa (Reuters, 2018). Since the government has not elaborated on how expropriation without compensation would work in practice, it is difficult to predict what impact this will have on mining. The next section provides an outline of the post-apartheid mining regulatory regime in South Africa.

Post-apartheid mining regulatory regimen

This section is not meant to be exhaustive, in any manner, and readers are encouraged to consult Cawood and Minnitt (1998) and Mitchell et al. (2012), for example, for a more elaborate historical perspective of mining legislation, mineral ownership rights, and mining and communities. As is widely known, the 1994 post-election period in South Africa heralded significant changes in all spheres of life, including mining policies and legislation. In a concerted attempt to redress the exploitative and discriminatory mining regulatory regime spanning decades of white domination, the ruling African National Congress (ANC) drew heavily on the 1955 Freedom Charter, the ANC-led alliance's main programme encapsulating historical black liberation ideals (Suttner & Cronin, 1986; Rametse, 2016). The Freedom Charter is the ANC's blueprint for a non-racial South African society and, broadly, triggered a paradigm shift in thinking about the democratic rights of black South Africans and their protection under the law (Gottschalk, 2015).

Three key provisions of the Freedom Charter are relevant to the discussion of land and the extractive industry in South Africa. These provisions begin with the preamble which declares the ownership of the country to "all who live in it," including "that no government can justly claim authority unless it is based on the will of all the people." The second provision declared, "The land shall be shared by those who work it," while the third proclaimed, "The mineral wealth beneath the soil, the banks and the monopoly industry shall be transferred to the people as a whole" (ANC, 1977; Ramatji, 2013; ANC, 2015).

Against the backdrop of the Freedom Chapter – and as expected – the first post-apartheid democratic government had a clear mandate to redress the inequalities of the past, including mineral rights ownership (Cawood & Minnitt, 1998), and more importantly in the context of this book, the protection of community interests (Mitchell et al., 2012). This mandate of the ANC is embodied in the Constitution of the Republic of South Africa, which enshrines the right of all South Africans to equality. Specifically, Section 9(2) of the Bill of Rights provides for specific measures to be taken to redress historical imbalances (Jackson et al., 2015).

These measures included Broad-Based Black Economic Empowerment (BBBEE), which has of late become a highly contentious issue (Jackson et al., 2015). The BEE Commission Report (2001: 1) made a bold statement in its opening page where it traces the origins of BBBEE to the "domination of business activities by white business and the exclusion of black people and women from the mainstream of economic activity." The BEE Commission Report (2001) provided the first formal definition of BBBEE, which is defined as:

> An integrated and coherent socio-economic process is aimed at redressing the imbalances of the past by seeking to substantially and equitably transfer and confer the ownership, management and control of South Africa's financial and economic resources to the majority of its citizens. It seeks to ensure broader and meaningful participation in the economy by black people in order to achieve sustainable development and prosperity.
>
> (BEE Commission Report, 2001: 2)

A year later, the Mineral and Petroleum Resources Development Act (MPRDA) 28 of 2002 was enacted to expand opportunities for historically disadvantaged people to enter the mineral and petroleum industries and to benefit from the exploitation of these resources (Department of Mineral Resources, DMR, 2002). The introduction of the MPRDA constituted a revolution in South Africa's mineral regime (Ramatji, 2013) as it terminated the common law concept of the ownership of mineral rights and vested the ownership of minerals in the country in the state, which merely regulates the exploitation thereof as a custodian of South Africans (Mitchell et al., 2012).

Additionally, the Act provides for compensation to the government for the country's permanent loss of the non-renewable resources by mining companies. Moreover, as far as revenue for the State is concerned, the MPRDA made provision that any extractor who acquires a mineral resource from within South Africa and subsequently transfers the right is liable to pay a royalty. In this regard, Section 3 of the MPRDA states, among other things, that

> as the custodian of the nation's mineral and petroleum resources, the State, acting through the Minister may, in consultation with the Minister of Finance, determine and levy, any fee or consideration payable in terms of any relevant Act of Parliament.
>
> (DMR, 2002)

The South African Revenue Service collects the royalties for these resources in terms of the Mineral and Petroleum Resources Royalty Act, 2008 and the Mineral and Petroleum Resources Royalty (Administration) Act, 2008. The MPRDA and other mechanisms, including the Mining Charters (a government document setting out black economic empowerment (BEE) targets and a blueprint for the transformation of the mining industry), are tools that the DMR has adopted to ensure that the holders of mining rights are committed to the development of the community through their various undertakings. Thus, holders of mining rights are required to submit a prescribed annual report, detailing the extent of compliance with the provisions related to corporate social responsibility (CSR) towards historically disadvantaged persons (DMR, 2002) (Table 3.2).

Table 3.2 Overview of post-apartheid mining legislation in South Africa

Year	Act	Purpose
1994	Mineral and Energy Laws Amendment Act 47 of 1994	To provide for the rationalisation of specific laws relating to mineral and energy affairs that remained in force in various areas of the national territory of the Republic by Section 229 of the Constitution, and to provide for matters connected in addition to that
2002	Mineral and Petroleum Resources Development Act 28 of 2002	To make provision for equitable access to and sustainable development of the nation's mineral and petroleum resources, and to provide for matters connected in addition to that
2003	Mining Titles Registration Amendment Act 24 of 2003	To amend the Mining Titles Registration Act, to substitute, add, or delete certain definitions; to re-regulate the registration of mineral and petroleum titles and other rights connected therewith and certain other deeds and documents, among others
2005	Precious Metals Act 37 of 2005	To provide for the acquisition, possession, smelting, refining, beneficiation, use, and disposal of precious metals; and to provide for matters connected therewith
2008	Mineral and Petroleum Royalty Act 28 of 2008	To impose a royalty on the transfer of mineral resources and to provide for matters connected in addition to that
2014	Changes to the South African Mineral and Petroleum Resources Royalty Act, 2008 (the "Royalty Act")	The purpose of the Royalty Act is to impose a royalty on the value of the minerals obtained from the earth and not on the value of the mineral derived from further beneficiation processes. The amendment saw the "minimum" requirement being deleted from Section 6A(1), as well as from Schedule 2 to the Royalty Act, in an attempt to provide clarity on how to interpret Schedule 2 when determining gross sales for, among others, unrefined minerals extracted beyond the condition specified for that mineral when it is transferred

Source: South African government (no date).

In 2014, further amendments to the MPRDA were passed (Roeder, 2016). As the development of legislation, including that related to mining, in South Africa during the post-apartheid era has been intertwined with the broader responsibility of meeting the rising expectations of citizens (Harrison, 2016), there was a need to demonstrate the government's commitment to democratic accountability. Changes to the MPRDA of 2002, which recognises the need to promote development and social upliftment of mining communities and to bring about more equitable access to South Africa's mineral resources reflect "citizen's spaces to claim accountability" (Bjuremalm et al., 2014: 19) insofar as their rights to be consulted about applications for mining licences pertaining to their land.

In adopting a more rigorous community-oriented approach, the MPRDA 2008 removed anomalies by providing several measures that enable communities to protect their interests in land that is the subject of applications for rights. The six measures relate to consultation, social and labour plans, preferent rights afforded to communities, and, as discussed earlier, the continuation of royalty payments in certain circumstances and BEE (Mitchell et al. 2012; Mostert et al. 2016). One of the broadest and far-ranging changes generated by the MPRDA 2008 relates to the definition of "community" by making an explicit reference to the requirement to consult (Mitchell et al. 2012: 112):

> A group of historically disadvantaged persons with interest or rights in a particular area of land on which the members have or exercise communal rights in terms of an agreement, custom or law: Provided that, whereas consequence of the provisions of this act, negotiations or consultations with the community is required, the community shall include the members or part of the community directly affect by mining on land occupied by such members or part of the community.

Many explanations have been advanced to account for why amendments were passed to the MPRDA, despite warnings from mining companies about its crippling effect. Johnson (2015: 155) summarised these concerns as including "strengthening the requirements placed on companies and greatly increasing the discretionary powers of the minister for mineral resources." In what I call the "brown envelope" phenomenon (polite elite taking kickbacks for state tenders or licences), Johnson (2015) states that the MPRDA has created a situation in which the Minister of Mineral Resources grants prospecting rights to favoured companies.

I shall return to a discussion of community engagement in a later section. In the subsequent section, I provide an outline of international legal and regulatory frameworks relating to community engagement, which has been the focal point for various international, national, regional, and local stakeholders.

Legal and regulatory frameworks for community engagement in the global mining industry – a mounting volatility

While mineral policy frameworks globally reflect the development trajectories of each nation, they are also a microcosm of international safeguard standards extending to national legislation. As Vivoda and Kemp (2017) observed, the codification and alignment of international safeguard standards and procedures into national law and regulation will in all likelihood influence the relationship between governments, mining companies, and local citizens. Thus, governments face considerable pressure to ensure that national policy frameworks in the mining sector expand local employment opportunities, increase tax revenues, and meet increasing community demands for improved infrastructure and more excellent environmental protection. As a result, mining companies face obstacles to investment, ranging from high royalty rates, permitting challenges, and uncertain tax rules to growing requirements for local beneficiation (Lane, 2018). Moreover, still, even though the mining industry has taken significant strides to improve its image in recent years, in some cases, the industry is operating with a legacy of weak environmental practices and fractious community relations (Ferguson, 2018). In this regard, the 2018 edition of *Tracking the Trends* contends that the imperatives of mining are playing out in interesting ways. In this regard, Deloitte (2018) notes that the challenges facing the mining sector have been continuing. During this time, the mining industry witnessed the emergence of innovative companies adopting transformative practices. The next ten years will see the continuation of rapid change in the industry. As a result, mining companies must rethink the traditional mining model.

The International Labour Organization (ILO) Convention 169, established in 1989, is the only international law designed to protect tribal people's land ownership rights and sets a series of minimum United Nations (UN) standards regarding consultation and consent about projects that affect tribal people. To date, only 22 countries have ratified the legally binding Convention 169 compared to the equally important UN Declaration on the Rights of Indigenous Peoples (Vivoda & Kemp, 2017). Many countries have thus absolved themselves from principles of community engagement as codified by international-level instruments. We are compelled, then, to ask profound questions about the extent of transparency in the management of extractive resources, and the willingness of governments to exercise effectively their responsibilities to oversee activities within the extractive industry. In essence, these flawed arrangements enable politicians to manipulate their countries' policies for their own benefit. This situation, coupled with a more excellent range of media sources and heightened community distrust of governments and mining companies, has led to a dramatic increase of civil society activism in mining communities.

To understand the legal and regulatory framework of the mining industry in Africa, one should not look further than the social responsibility of mining companies. The mineral regulatory regimen in many African countries imposes extensive social obligations on mining companies. In fact, the significant changes to mining regulations seen recently in various African states have given rise to a concern that a regional trend of resource nationalism may be re-emerging. These significant changes ranged from the drastic amendments to the 2010 Mining Act in July 2017 and two new laws asserting the Tanzanian government's "permanent sovereignty" over its natural resources to the 2017 Mining (State Participation)

Regulations in Kenya. The latter affords the government a free-carried 10% equity interest in any mining right granted – as in the case of Ghana – after the 2016 Mining Act has come into force on 27 May 2016 (Leon et al., 2018). Similarly, in Equatorial Guinea, the government claims ownership of all mineral resources, whereas The Mining Code (2014) in Côte D'Ivoire requires mining companies to prepare a development plan and to establish a fund for socio-economic development plans (Vivoda & Kemp, 2017).

Another characterisation of mining regulation in Africa (and elsewhere in the world, for example, Italy, Portugal, Spain, and Switzerland) relates to land tenure, when customary and statutory legal systems overlap. Customary tenure refers to collectively owned land, usually under the authority of a traditional leadership (Berry, 2017) or large families or clans (Comaroff & Comaroff, 2018), as in the case of Papua New Guinea. In this regard, the 2009 Framework and Guidance on Land Policy in Africa of the African Union (AU) advocates a leading role for customary tenure in land governance (Chimhowu, 2019).

In countries such as Burkina Faso, Rwanda, Tanzania, and Uganda, the formalised customary tenure allows for the registration of individual and collective titles. There are also several countries in Africa where customary law dictates that land cannot be alienated or sold. Customary law in countries like Côte D'Ivoire allows the right to use the soil to be transferred or sold, but not the ownership (Vivoda & Kemp, 2017).

There has been a paradigm shift, however, in how customary land is governed, which Chimhowu (2019) referred to as the neo-liberalisation of customary tenure in Sub-Saharan Africa. In Ghana, for example, Article 20 of the Constitution of 1992 grants the State the authority to acquire private land in the public interest. Interestingly, though, approximately 80% of Ghana's land is owned by customary authorities. The customary land tenure framework places chiefs – the custodians of traditional communities – in a pole position when it comes to securing access to surface land (Vivoda & Kemp, 2017). However, there is a significant lack of community engagement and consent requirements in Ghana and Tanzania, which mirrors the position in South Africa (Mandela Institute, 2017).

In an attempt to boost investor confidence, the Indigenisation Act in Zimbabwe previously required that at least 51% of all public companies and businesses had to be owned by indigenous Zimbabweans. However, the Indigenisation Act now only applies to companies involved in the extraction of diamonds or platinum (Becker, 2018). Historically, indigenisation has been criticised for deterring foreign investment, promoting neo-patrimonial politics, and encouraging elite wealth capture. Other post-independence economic reforms include the first indigenisation policy, the Fast Track Land Reform (FTLR), which was blamed for the collapse of the agricultural industry and the subsequent meltdown of the Zimbabwean economy (Gilberthorpe, 2015). Acceptance of indigenisation plans in Zimbabwe depends largely upon ministerial approval of who is to benefit. The regulations give the Minister of Indigenisation and Economic Empowerment a mostly unfettered discretion to decide whether to approve or reject an indigenisation plan or to attach conditions to such a plan (Matyszak & Research and Advocacy Unit, 2011). The complexity of defining "indigenous peoples" and "previously disadvantaged communities" makes it difficult to craft laws that are exclusively aimed at benefitting the target groups. Often, it leads to opportunists hijacking opportunities meant for the local communities. In South Africa, for instance, there is a general perception that the BEE laws have benefited mainly the upper-middle-class elites; the same fate is likely to befall Zimbabwe (Murombo, 2013). Huni (2017) recommends that there is a need to draw up a new mining sector policy – guided by sound economic principles and not politics – and take into account the shortcomings of the beneficiation-enabling environment of Zimbabwe.

Perhaps, the most exciting feature of mineral regulatory frameworks in Africa is the Africa Mining Vision (AMV), which is a policy framework that was created by the AU in 2009 to ensure that Africa utilises its mineral resources strategically for broad-based inclusive development. The implementation of this transformative policy, which supposedly can drive sustainable development on the African continent, has been lethargic (Economic Commission for Africa, 2011). Views expressed at the Alternative Indaba of Mining in 2018 contend that Africa has reverted to the pre-independence days of colonialism in which legal regimes exist primarily to facilitate profit for mine owners. Historically, it is argued, the majority of African countries opted to nationalise mineral resources at independence, a choice that vested the ownership of mines in the state on behalf of the people of that country. However, the current legal regimes in Africa primarily facilitate the profits of mining companies but not how best the interests of the people were advanced (Acca, 2018).

Although the extensive reforms of regulatory and legal frameworks introduced during the 1980s and 1990s helped to create a more favourable environment for foreign investment in African mining, their contribution to social and economic development objectives has been far less individual. As a result, active and vibrant civil society movements, protesting about the costs and questioning the benefits of the mining sectors, have emerged in many mineral-rich African countries (Economic Commission for Africa, 2011).

Elsewhere in the world, Australia notably has adopted a federal legislative framework for the mineral extraction process and is divided throughout the three levels of government (Hunter, 2015) and all vesting provisions are derived from state and territory legislation (Hepburn, 2015). Native title rights and the protection of the cultural heritage of Aboriginal and Torres Strait Islander people are enshrined in the NSW Australian government's mining regulatory regimen and other federal states. Generally, the Australian mining regulatory framework regards community engagement as an essential element in the planning and decision-making processes of the mining industry.

In Mexico, the mining regulatory framework was profoundly changed after international organisations such as the World Bank, the International Monetary Fund, and the Inter-American Development Bank demanded that it adopt more neoliberal policies to open the country's weak economy to foreign investment in the 1980s. These market revisions in 1992 enabled the North American Free Trade Agreement (NAFTA) to allow foreign companies the opportunity to ignore national legislation concerning social rights. Furthermore, NAFTA does not accept complaints from local inhabitants or communities (Stoltenborg & Boelens, 2016). In countries like Chile, Papua New Guinea, and Peru, where indigenous people cannot be dispossessed of their land without state approval, the principles of NAFTA will be met with opposition.

In Canada – a stable democracy with a well-established rule of law and extraordinarily well endowed with natural resources – private land ownership does not extend automatically to the ownership of sub-surface minerals, as in the case of the United States. In effect, primary resource ownership was assigned to the federal government and ten provincial governments, with each province having its own set of laws and policies dealing with mining (Grant et al., 2014). In Ontario, for example, the government's initiatives were directed towards establishing a stable legal environment by promoting the legal rights of prospecting and mining companies. As a common practice in much of the world, the Canadian government and mining industry were the key role-players in the mining industry.

In some instances, the national mineral policies are shaped by the level of resource wealth. Mineral resource-poor countries requiring substantial mineral inputs, such as Japan, the Republic of Korea, and Taiwan Province of China, emphasise different policy objectives than mineral-rich non-industrialised nations, such as Papua New Guinea, New Caledonia, or the Democratic Republic of the Congo. Insofar as the shale gas industry in Eastern Europe is concerned, Goldthau (2018) provides an instructive account of the distinct way governments interact with private and social actors, and how these interactions are structured through institutional setting processes.

Globally, there has been a shift in recent times in the principles and practices of community engagement in the mining industry. Leading in this regard is the International Council on Metals and Minerals (ICMM), which committed its members to a free, prior, and consent (FPIC) process. The notion of FPIC is underpinned by several principles, including adequately informing local communities about projects and meaningful engagement that reaches beyond the social licence to operate (Oxfam, 2015). There has also been a high degree of policy convergence on performance standards related to community engagement from the Sustainability Framework of the World Bank's International Financial Corporation, the Organisation for Economic Cooperation and Development (OECD) Guidelines on Multinational Enterprises, and the UN Guiding Principles on Business and Human Rights (Mandela Institute, 2017). For the most part, the focus of these international instruments is to monitor corporate and state commitment to community engagement. Lately, however, there has been a concerted move in mining regulatory frameworks to allow for local-level agreement-making, which is discussed in the South African context, in the next section.

LLAs in the mining industry

Local-level agreements (LLAs) – a generalised term, which includes terms such as community development agreements (CDAs), Indigenous Land Use Agreements (ILUAs), and Impact Benefit Agreements (Harvey, 2018) – with mining communities throughout the different stages of the operations of a mine are fundamental to successful resource development. These LLAs are an affirmation by mining companies of the significance of operating by a social licence. In countries, for example, Mongolia, where mining has sparked conflict between mining companies, the government, and mining communities, LLAs have become an essential means to secure an SLO (Dalaibuyan, 2015).

Two distinctive features of the Minerals Law of Mongolia are, first, the mandatory nature of LLAs and, second, the fundamental and direct role assigned to local governments in agreement-making (Dalaibuyan, 2017). Additionally, Mongolia is one of the 52 countries that have adopted the Extractive Industries Transparency Initiative (EITI) – a global standard to promote the open and accountable management of extractive resources. The EITI standard "requires the disclosure of information along the extractive industry value chain from the point of extraction, to how revenues make their way through the government, and how they benefit the public" (EITI, 2019).

The rapidly expanding footprint of mining has been met with an equal spread of LLAs across all regions of the world, albeit in widely varying contexts, scope, and content. For example, in Australia, Ghana, Nigeria, Greenland, Papua New Guinea, and Canada, LLAs are required by law (Gathii & Odumosu-Ayanu, 2015; Harvey, 2018) and are voluntary in Argentina (Haslam, 2018). Notwithstanding these differences, the content of LLAs generally includes agreement on what constitutes local, governance complaints, dispute-resolution procedures, and the establishment of dedicated trusts to receive funds for purposes specified in the LLA (Harvey, 2018). Above all, Gathii and Odumosu-Ayanu (2015: 69) argued that these contractual forms demonstrate that the law of contract has evolved from the

> nineteenth-century idea that contracts merely protect the rights of investors without much concern for those who are directly affected by extractive industry operations. By including affected communities, indigenous communities, and others, these new contractual forms demonstrate that investors and governments are trustees and that extractive resources must be mobilized for the benefits of their publics.

Harvey (2018) identified two principles around which successful LLAs revolve. First, LLAs should centre on mine–community social groups and their representatives, rather than governments. Irrespective of governmental presence, it is crucial that there is an institutionalised relationship between a local social group and the miner, rather than something agreed between certain individuals. Second, some business principles are also key to an LLA's success. For example, the way in which revenue streams will be managed and allocated (O'Faircheallaigh, 2012), and a mutual obligation and agreement that all parties will commit to shared objectives must be set out. In sum, the expected outcomes of the LLA should be clearly described, with measurable performance indicators and sanctions for non-performance (Harvey, 2018).

Practical challenges of CDTs

In Chapter 1, I outlined how CDTs are used as informal and voluntarily initiated LLA implementation units, and a means to secure an SLO. The literature has given scant attention to the issue of CDTs in South Africa. Thus, insights about the failures and successes of CDTs are filed in company archives, which deny stakeholders the facility of constructing lessons from experiences. The trusts are regulated by the Trust Property Control Act (Act No. 57 of 1988) as amended. Community trusts have their pitfalls, but if trustees that are fully committed to good governance are appointed, communities will benefit through job creation and the mobilisation of resources (Nelwamondo, 2016).

Some CDTs regarded as successful in South Africa include the Royal Bafokeng National Development Trust and the Sishen Iron Ore Company (SIOC) CDT. SIOC is the super trust and has several local trusts, such as the Gamagara Development Trust, Tsantsabane Community Trust, and the John Taola Gaetswewe Development Trust (Nelwamondo, 2016). The primary goal of SIOC established in 2006 by Kumba Iron Ore Limited, for example, is to invest in the development of the communities in which the company operates in the Northern Cape and Limpopo (SIOC, 2019).

Historically, though, corporate social investment and CSR are not new in South Africa. In the mining sector, companies have been ploughing money into education, housing, and other areas long before the advent of democracy. For example, the Anglo American Chairman's Fund – one of the largest donors in the country – dates back to the late 1950s and its new funding priorities go far beyond mining communities. These priorities include early childhood development; maths, science, and language development at school level; strengthening the public health-care system; and capacity building for community-based and other non-governmental organisations (Kane-Berman, 2017). Anglo American (2019) states on its website that the initiatives contribute directly to progressing SLP commitments, which are aligned to municipal integrated development plans (IDPs). Anglo American further reports that each of their operations has an SLP, which works correctly towards establishing and improving health and welfare, education and infrastructure within our mining communities. These plans are developed through a consultative process with local municipalities as well as through regular interaction with host communities to ensure that the identified projects are sustainable and in line with their needs. Anglo American (2019) further notes that "these initiatives are broader than just our SLPs, and we have enjoyed a number of enterprise development and CSI [corporate social investment] successes, including some projects that would normally be the responsibility of municipalities."

Seeking to upset existing empowerment mechanisms initiated by the Department of Trade and Industry, which primarily enrich a select few through their empowerment schemes, companies in various sectors, including mining, established community trusts in which benefits reach a broader base. Despite their challenges, CDTs remain the best model for communities to mobilise resources and create jobs, and their success is based on how they are structured at the outset. A trust's deed and a clear definition of its objectives are critical. Trustees of these organisations have a relatively complex and demanding set of tasks, ranging from communicating with beneficiaries and identifying projects to issues of corporate governance. Appointing the right people is central to a community trust's success, but finding suitable trustees is not always easy, particularly in rural areas where literacy and education levels are low. Stimulating interest and spreading understanding about the community trust and the role of trustees might help to attract suitable candidates, such as teachers, nurses, or doctors, who are educated and who have the community's interests at heart (Nelwamondo, 2016).

However, setting up useful, representative trusts that allow for real community participation is fraught with difficulty. CDTs risk being hijacked by local government and community leaders looking to advance their own positions. Other influential local stakeholders, including chiefs and ward councillors, also have their own priorities, which can make them problematic as trustees. There are often multiple parties affected by or involved with the trust or its projects, including traditional leaders, traditional councils, municipalities, community-based organisations, non-governmental organisations, and communal property associations. Furthermore, while CDTs are supposed to be non-partisan, the voiceless within the beneficiary community tend to remain voiceless. Due to limitations on available resources for operational expenditure, a community trust is often unable to conduct active engagement with intended beneficiaries (Sidley, 2015).

The CDTs therefore reflect South Africa's version of LLAs as well as the significant business considerations inherent in a company's approach to CSR. However, the governance of mining development in South Africa has remained business-centric, with the government viewing its regulatory decisions as a mandate that can be imposed on its citizens (Truter, 2011). Scholars argue that this mandate has the potential to encourage lower rates of compliance with fundamental measures of an SLO, such as local accountability and local ownership (Matebesi & Marais, 2018). Given the imperative of mining companies to work in harmony with local communities (Sander, 2018), the increasing shift towards an era of social comprises an important challenge to the traditional way governments and mining companies have engaged with mining communities. Buoyed by the expanded protection against the environmental damage provided by mining laws, and the heightened political consciousness stirred by anti-extractive movements, civil society forces encouraged renewed, widespread community struggles against mining companies.

Civil society mobilisation against mining

Globally, anti-extractive movements can stall or disrupt mining projects. Already, we have witnessed the growth of more extensive popular movements against mining (Riofrancos, 2019). For example, the Bodo community in the oil-producing Niger Delta, which collected baseline data to prove that two Shell oil spills caused significant environmental damage (Pegg & Zabbey, 2013).

A British court ruled in 2018 that the Bodo community should retain the option of litigation until 1 July 2019 (Africa News, 2018). In Bolivia, the Tacana community took control in negotiating an environmental impact assessment (EIA) when a Bolivian oil company wanted to drill on its land (Looby, 2017).

In what perhaps serves as one of the tragic global cases of state repression against community, mobilisation against mining is the clampdown of the Ecuador government against indigenous Shuar activists. The years of violent clashes between security forces and community activists at the site of a copper mine left three Shuar activists and one policeman dead. This led to the declaration of a state of emergency which lasted three months, but the harassment of Shuar activists by state agents continues (Riofrancos, 2019).

In recent times, civil society struggles against mining policy and mining in general have gained renewed interest. Abuya (2016) identified six core issues that mining conflicts revolve around. These mining struggles have been driven by six core issues (Abuya, 2016). These issues are land ownership, unfair compensational practices, inequitable resource distribution, environmental degradation, mine-induced poverty, and human rights abuses. Missing from Abuya's (2016) identification of core issues that lead to conflict in mining is community engagement.

Over the years, mining companies have used several strategies to minimise company–community conflicts. These strategies or company motivations vary along a continuum described by Banks et al. (2017) as ranging from contractual obligations to corporate philanthropy, while considerations of CSR and an SLO are somewhere in the middle. Mining companies, Banks et al. (2017: 215) argued, "bring weighty global charters and lofty community-focused corporate rhetoric to their local programs." CSR-driven initiatives have been one of the most common strategies used by mining companies to minimise conflict with mining communities. However, scholars argue that much conflict between mining communities and mining companies is fostered by the unilateral power of mining companies to pursue CSR programmes in ways that advance company interests (Abuya, 2016; Banks et al., 2017). When mining companies realised the limitations of CSR programmes, the overall trend has been a move towards the "corporatization of activism," where the agenda, discourse, questions, and proposed solutions increasingly conform with, rather than challenge, the status quo (Dauvergne & LeBaron, 2014: 4).

Over time, though, CSR-driven initiatives proved inadequate to appease the widespread discontent that has grown in the wake of the persistent failure of mining companies and governments to consult with mining communities about mining projects in their areas. As a result, community mobilisation against mining has increased dramatically in terms of scope and intensity in many parts of the world, including South Africa.

The context of civil society mobilisation against mining in South Africa

The context in which mining policy is devised in South Africa is a mixture of participatory and representative democracy, which allows citizens to elect their public representatives at different levels of government. There has been much debate over the years about how civil and socio-economic rights, espoused by the South African Constitution, translate into the actual dynamics of participation, which are meaningful and empowering to citizens (Modise, 2017). Nonetheless, South Africa has a long history of a vibrant and active civil society. While governments elsewhere in the world (for example, Bolivia, Ethiopia, India, and Bangladesh) are intensifying efforts to suppress the ability of civil society, the culture of civil society activism in South Africa is as vibrant as ever.

Traditionally, labour unrest by organised labour has been one of the most prominent challenges that has increased risk in mining in South Africa. The high prevalence of labour unrest in the mining industry – primarily waged over increased wages, housing and employment conditions – has turned more violent as a result of union rivalry. These protests often led to the suspensions of mining operations to over three with significant impact on the South African economy, mining production, investor confidence, and foreign direct investment. A case in point is the unprotected labour strike at Lonmin mining company's Marikana mine in the North West Province of South Africa, where 34 striking mineworkers were shot dead by the police in August 2012 (Hill & Maroun, 2015).

With respect to conflicts between mining companies and mining-affected communities, the literature reveals that South African mining laws do not require the government to share mining agreements with the public. In this sense, the MPRDA fails to achieve its stated goals in respect of empowering communities. For example, since SLP regulations do not require public disclosure of the plan, the beneficiary communities, and other interested stakeholders. The Mandela Institute (2017) argued that since SLPs are a licensing condition, the lack of disclosure by mining companies creates a level of distrust between the excluded communities and interested stakeholders and mining companies. Another controversial stipulation of the MPRDA is the emphasis on community in the development of SLPs, but this consultation is not mandatory. The Mandela Institute (2017: 21) concluded:

> Further criticism surrounds both legal provisions regarding community engagement and how this operates in practice. The regulatory scheme in South Africa 'presupposes a one-size-fits-all model for such communities, despite their diverse needs and circumstances, and reserves no seat at the regulatory table for the affected mine communities.' This is related to and compounds difficulties defining 'indigenous peoples' and 'previously disadvantaged communities' in the South African context, which makes it difficult to design laws that effectively benefit these groups. On the whole, these problems with the South African mining context, laws and their implementation disempower local communities. Taken together, the lack of information, weak legal protections and the difficulties of establishing mechanisms to ensure that communities benefit and participate, make it hard for communities to hold the government and mining companies accountable and pursue protection of their rights and interests.

CSO response to exclusion to mining agreements

Over the past decade, the bureaucracy that deals with mining in South Africa has been characterised by the striking of deals behind closed doors between the DMR and the Minerals Council of South Africa (Minerals Council: formerly the Chamber of Mines), which is a representative organisation of mining employees. A typical example of the exclusion of mining and affected communities has been the tension between the government and the Chamber of Mines over the failure of mining right holders to meet the full requirements for meaningful participation. In May 2015, the Chamber of Mines boldly claimed that the "mining industry has done more for transformation than any other component of the private sector since the advent of democracy" (Nicolson, 2015).

However, in a report assessing industry compliance with the 2004 Charter and its 2010 amendment, the then Minister of DMR, Ngoako Ramatlhodi, announced that 45% of mining companies did not meet targets for improving living conditions for miners and only 36% met targets for mine community development (Nicolson, 2015). Rutledge (2015) argued that the stance of the Chamber of Mines did not reflect the lived reality of communities affected by mining. He further asserted that the divergent views of the minister of mineral resources and the chamber did not obscure the fact that they are both "united in their exclusion of the people they supposedly care so much about."

Meanwhile, an extensive popular movement emerged in South Africa demanding a just and equitable mining industry and against attempts to further the narrow interests of the mining industry and the government. As a result, Mining Communities United in Action (MACUA), a national organisation established in 2012 representing over 70 organisations and 20 CSOs, met in mid-2015 to consider the continued exclusion of communities from mining laws. This effort by civic organisations was also aimed at the failure of the government and the mining industry to conform to a constitutional imperative imposed on them to consult with those directly affected by mining (Rutledge, 2015).

These developments bolstered the vibrancy and visibility of civic activism against mining in South Africa. A year later, in 2016, seven CSOs, which included the Centre for Environmental Rights, the Highveld Environmental Justice Network, and Earthjustice, made a joint submission to the UN Human Rights Council, blaming South Africa's inadequate regulation of the mining industry and coal-fired power stations for the violation of human rights in the country. A significant part of the report focused on recommendations that mining companies be held accountable for unlawful activities through a comprehensive and transparent compliance and enforcement programme. Criticism was also directed at government inaction, which, the CSOs argued, had made the mining industry one of the least transparent industries in South Africa (Faku, 2016).

Mining charters and court challenges by CSOs

Broadly, no issue has generated so much conflict between CSOs and the government than concerns around the mining charters. In early 2016, the government released the Draft Reviewed Mining Charter, which brought changes to the mining charter promulgated in 2010. At the time, the Chamber of Mines viewed the draft mining charter as containing ill-considered and unrealistic targets.

In respect of communities, the proposed charter stipulated that traditional authorities significantly increased the target for black representation at the management level (Baxter, 2017). The new mining charter has several other shortcomings for communities. These shortcomings include, for example, not adequately addressing the drastic imbalance between local ownership and foreign ownership. Second, it ignores the negative impact that mining has on mining communities. Third, the 5% community share is not related to the value of the minerals being extracted and the community stands to lose its land to mining and will never recover. Fourth, communities will not be able to afford the proposed BBBEE shares, which are sold to BBBEE shareholders (Bench Marks Foundation, 2016).

Interestingly, it would be that an urgent court interdict of the Mineral Resources Council sought to prevent the then Minister of Mineral Resources, Mosebenzi Zwane, from implementing the 2017 Mining Charter that would provide a catalyst to CSO activism against exclusion in granting an SLO. This court interdict followed multiple court challenges by the industry in 2017 (Moalusi & Malesa, 2019). On 14 November of the same year, the Pretoria High Court ruled that mining community networks be granted leave to intervene in the review of the Mining Charter. This court ruling essentially gave communities a voice in the drafting and implementation of mining charters (Bruce, 2017).

The next year was the most eventful in respect of court rulings in favour of mining communities and CSOs. The year 2018 yielded a new minister of mineral resources when the newly appointed president of the ruling ANC replaced Mosenzi Zwane with Gwede Mantashe. The Chamber of Mines welcomed the arrival of Minister Mantashe, whom they hailed as a man of integrity and dignity (Groenewald, 2018).

The euphoria of Mantashe's arrival as Minister of Mineral Resources did not extend to mining and affected communities as the unilateral actions of the Ministry of Mineral Resources and the mining industry continued unabated. Consider the contributions of Ramphele (2019) in this regard. She argued that the confrontational stance of Gwede Mantashe often obscured more than it illuminated and that the efforts of civil society pressure groups to engage him were futile. This prompted Ramphele (2019) to state that "it seems that Gwede Mantashe is not paying sufficient attention to the core principles of constitutional democracy that puts citizens at the centre of shaping their futures."

The power of civil society was demonstrated in part by the successful court challenges launched by CSOs against what they termed the "persistent exclusion from the iteration of the mining charter," as well as mining rights. In early 2018, MACUA and other concerned civil society actors started an online petition, aptly called the "Right to Say No," which was based on the FPIC principle. The goal of the petition was to support the Xolobeni community on the Wild Coast in the Eastern Cape Province of South Africa, which was in court to fight a petition to reject an application by an Australian mining company to mine titanium in its community. A central idea underlying the petition led by MACUA was three-pronged. First, it aimed to demonstrate to the court that communities had a fundamental right to be involved and informed about developments that affect them. Second, it aimed to allow communities to say "no" to cases of unsatisfactory outcomes of negotiating processes. Third, MACUA's plan was also for the court to issue a declaration order that forbids DMR from issuing a mining licence against the will of the Xolobeni (Mabuza, 2018).

In November 2018, the North Gauteng High Court in Pretoria ruled in favour of the Xolobeni community in a historic mining rights case in South Africa (Mitchley, 2018). The judge declared that the Mineral Resources minister must obtain consent from the community as the holders of the rights on land, before granting any mining right to the Australian mining company, Transworld Energy Minerals (TEM). This landmark judgement, with a profound impact on the prominence of community engagement, acceptance, and approval of mining projects, underscored the significance of the Interim Protection of Informal Land Rights Act (IPILRA). The IPILRA, enacted in 1996, to secure land rights to South Africans living in former homelands, provides that people may not be deprived of their natural land rights without their consent, except by expropriation (Claassens, 2019).

The Xolobeni judgement followed another one a month earlier in October 2018, dubbed by Mabuza (2018) as a "massive victory for small community in mining rights case." The case relates to the Lesetlheng community in the North West Province of South Africa, which challenged an eviction notice, previously approved by the North West High Court. The interdict and eviction order brought by Itireleng Bakgatla Mineral Resources (IBMR) and Pilanesberg Platinum Mines (PPM) was an attempt to prevent the Lesetlheng community from continuing to cultivate and keep their cattle on the platinum-rich farm, which they have farmed for almost a century (Cabe, 2018). The Constitutional Court set aside both the interdict and eviction order that was granted in 2017, and thus reaffirmed that having a mining right does not surpass the rights of those who currently occupy the land. In delivering his written judgement, the judge quoted extensively from revolutionary Caribbean thinker Frantz Fanon's *Wretched of the Earth* (1963: 9):

> For the colonised people, the most essential value, because the most concrete, is first and foremost the land: the land, which will provide bread and, above all, dignity ... [The judge continued] Thus, strip someone of their source of livelihood, and you strip them of their dignity too.

Sustained community mobilisation in Xolobeni and Lesetlheng

Xolobeni and Lesetlheng are two communities that are about 1,000 km apart, yet share several common challenges faced by poor communities in mineral-rich former homelands. The communities represent an innovative form of civic engagement in which communities take the lead to express their wish to be consulted and involved in decisions about developments that affect them. Both communities do not have formal title deeds to the land that they occupy, due to past discriminatory laws. The Lesetlheng community bought the land a century ago, and it was registered in the state's name, but held in a trust for the Bakgatla-ba-Kgafela tribe. The Xolobeni community never purchased the land (Claassens, 2019).

First of these challenges is routine dispossession without consultation or compensation. Due to racially discriminatory laws in South Africa, the land owned by the two communities had not been registered in their names but in the names of their traditional authorities. In this regard, mining companies often used the MPRDA and erroneously argued that the right to mine supersedes all other rights over the land, including the protection of surface rights enshrined in the IPILRA. The Department of Mineral Resources and the MPRDA continue to be inadequate. For example, Section 54 of the MPRDA provides for compensation to be determined only after a mining right has been granted, mining has commenced, or a dispute has arisen. In such cases, the intervention of the DMR is required. However, despite ample disputes arising between communities and their traditional leaders and mining companies, the department has never initiated a Section 54 dispute-resolution process (Claassens, 2019).

In Xolobeni, the Amadiba Crisis Committee (ACC), local community-based organisation, has been waging a 16-year-long struggle against mining in the Wild Coast of South Africa. The proposed titanium opencast Xolobeni mine – which would have been one of the largest of its kind in South Africa – by Australian mining company TEM, was first granted a mining licence in 2008. The licence was withdrawn three years later in 2011, due to the failure of the mining company to address several outstanding environmental issues identified by DMR at the time of application. Subsequently, the ACC withdrew the complaint that it had lodged with the Public Protector due to numerous delays by the government to decide on the matter (Van der Merwe, 2011).

Over the years, the Xolobeni villagers have been advancing that mining in the community would result in the removal of more than 70 households from the farming community and, thus, the disruption of the lives of inhabitants of the land. The community was in favour of sustainable livelihoods from nature and heritage-based eco-tourism. In September 2017, the DMR announced an 18-month moratorium on mining in the area (Moosa, 2018), with an extension proposed in August 2018 by Minister Mantashe. The long protracted struggle and legal battle of the ACC, which has resulted in intimidation and assassinations in the Xolobeni area, yielded a significant victory for mining communities in South Africa when the Amadiba community won a High Court case against the mineral resources minister and DMR in November 2018.

Despite having exercised exclusive use and control of the Wilgespruit farm for a century, 13 families in Lesetlheng, which were removed from their ancestral land to make way for the expanding platinum mining operations, challenged an eviction notice, previously approved by the North West High Court.

The interdict and eviction order brought by IBMR and PPM was an attempt to prevent the Lesetlheng community from continuing to cultivate and keep their cattle on the platinum-rich farm, which they had farmed for almost a century (Mabuza, 2018). The crisis facing the Lesetlheng community in their struggle to keep their ancestral land began when the DMR awarded a mining right to IBMR to mine platinum on the farm in May 2008. Soon after, the Minister of Mineral Resources and the Bakgatla Ba Kgafela Tribal Authority agreed, instead of the 13 affected families. Four years later, the exclusion of the 13 affected families continued when IBMR concluded an agreement in which PPM would mine part of the farm. PPM immediately began clearing vegetation and excavating soil. Full-scale mining operations began on the farm in 2014 (Booi, 2018).

Subsequently, the first significant opportunity for the Lesetlheng community to challenge the eviction order and an interdict granted in 2017 to the PPM to evict the community from Wilgespruit farm came when the community applied for a spoliation order. At the time the eviction order was granted, the court said the Lesetlheng community members were not actual owners of the land, as it was not registered in their names. The court further argued that the mining company informed the community that they now had mining rights to the land, and this should be construed as a consultation. The Lesetlheng community lost on appeal in the Supreme Court of Appeal and then petitioned the Constitutional Court, which led to the landmark ruling in which the court concluded that "the existence of a mineral right does not itself extinguish the rights of a landowner or any other occupier of the land in question" (Evans, 2018).

The challenges facing Xolobeni and Lesetlheng in their quest for recognition from the government and mining companies are illuminative but not unique. What started as grassroots forms of activism – supported by leading CSOs, social justice and legal experts – has become a national issue that forges a new vision for community mobilisation against mining. As a consequence, communities such as Xolobeni and Lesetlheng have been marginalised by the Minister of Mineral Resources due to their autonomous and critical stance, which aims, in part, to highlight their demand for an alternative economic model in place of the traditional extractive model of development. These alternative development models, Ramphele (2019) argued, include organic carbon-based agriculture indigenous to Xolobeni, ecotourism, livestock farming, and food production.

A case in point is the minister's attempts to ignore the High Court judgement that reaffirmed the rights of the Xolobeni community over their land on the basis that he suspects that there will be no mining if the right to issue licences is shifted to communities (Ramphele, 2019). Another example is the minister's narrow interpretation of the court judgement in respect of community consultation by announcing his plan to conduct an independent survey to gauge the opinion of residents, a move that divided the community (Mabe, 2019). Interestingly, the minister has decided to ignore the blatant abuse of power by traditional authorities: in both Xolobeni and Lesetlheng, the traditional leaders are shareholders in the mining companies that want to evict the communities from their land.

Additionally, a report – spurred by the Xolobeni community's resistance to the Australian mining company – has shown the extent to which the mine failed in its duties to assist people in the Matzikama Local Municipality (along the west coast in the Western Cape Province of South Africa). A South African subsidiary of the mine had the rights to mine 15 km of beach along the coast of the municipality when it began operations in 2015. By 2018, the mine had breached its legal boundaries leading to the collapse of a 17 m cliff. The government has yet to hold the mining company accountable (Pather, 2018).

This points to an underlying tension between traditional power and the rights of mining communities. This tension is exacerbated by the fact that traditional management models in South Africa assume the primacy of traditional authorities or leaders. Nevertheless, how can the South African government justify the failure to act upon the conflict of interests and blatant unethical conduct by some traditional leaders, however culturally significant their roles, without considering the circumstances of mining communities?

Ramphele provided, in part, the answer to the question in an opinion piece entitled "Xolobeni and the Irony of History" (Ramphele, 2019). In a direct attack based on Minister Mantashe's argument about communities taking away the right of the government to decide over mining licences, Ramphele (2019) noted that

> it seems that Gwede Mantashe is not paying sufficient attention to the core principles of constitutional democracy that puts citizens at the centre of shaping their future. She then asked: Was the struggle for freedom not to promote decision making by communities such as the Xolobeni community? She further lamented that it seems that the minister of mineral affairs is not paying sufficient attention to the core principles of constitutional democracy that puts citizens at the centre of shaping their futures with government officials as the servants of the people, not their masters. "One would have thought that Mantashe as a former migrant worker in the industry, leader of National Union of Mine Workers, former chair of the SACP, former ANC secretary general, and now chair of the ANC that styles itself as the leader of society, would demonstrate greater empathy for courageous communities such as the Amadiba, instead of resenting their determination to assert their rights."
>
> (Ramphele, 2019)

Ramphele's (2019) views bring to the fore the inescapable political dimension of community mobilisation for the right to take centre stage in the issuing of an SLO. The preceding discussions illustrate some of the different strategies of CSOs in advancing the rights of mining communities. Community mobilisation against mining companies and the government is embedded in the broader politics of party identity and activism in South Africa. For example, Matebesi and Botes (2017) noted that the connection between party affiliation and social movement is blurred in the country. In this regard, partisan protesters are consequently able to navigate successfully between the party and grassroots organisations. Matebesi and Botes

(2017: 94) emphasised that there is a fluid space occupied by partisan activists that help us understand the strategies used by protesters in South Africa.

What does the future hold for community mobilisation against mining in South Africa? The intricate strategies used by both the government and civil societies referred to as "actions-reactions-adjustments-readjustments" (Gupta, 2017: 240) point to the future redefining of activism against mining in the country. Booi (2018), for example, argued that as mining disputes have been making their ways through courts, the government has sought to enact a law that explicitly elevates the status of traditional leaders over rural people.

The continued marginalisation of precarious mining communities, it seems, will carry on after the passing of the controversial Traditional and Khoisan Leadership Bill. The bill, which empowers traditional councils to conclude mining agreements without obtaining the consent of communities, fundamentally compromises the property and citizenship rights of rural communities (Kiewit, 2019). For a response to the bill, one should, perhaps, not look further than the call made by Civil Society Coalition on the MPRDA in a joint media statement in 2015, which noted, "only inclusive solutions offer any hope of a stable and prosperous future. Shortcuts and authoritarian impositions are not the answer, but rather the most dangerous way forward" (Centre for Environmental Rights, 2015).

Conclusion

The chapter examined the evolution of a legislative and policy framework for mining, with specific emphasis on land rights and royalties, community engagement, community mobilisation, and local-level agreements in mining in South Africa. While the MPRDA constitutes a significant step towards breaking the logjam and changing racially discriminating laws that have caused harm to South African communities, it is evident that the formal recognition of these communities in the granting of an SLO is still elusive. This is a long-standing problem across the world and has become one of the primary reasons for the contentious relationship between mining communities and mining companies or governments.

Also, as Petras and Veltmeyer (2019: 202) eloquently put it, "the growing protest movement against mining capital and extractivism in the post-neoliberal era has engaged the forces of resistance not just against neoliberalism and globalization, but against the operative capitalist system." Thus, the so-called politics of natural-resource extraction is not merely a matter of better resource management, a post-neoliberal regulatory regime, a more socially inclusive development strategy, or a new form of governance – securing the participation of local communities and stakeholders in decisions and policies in which they have a vital interest.

In this regard, several insights are especially evident in this chapter in respect of civil society mobilisation for more excellent representation, participation, and the right to consent to mining operations by local communities in South Africa.

First, mining-related legislative and policy frameworks take place in the context of international standards and national political imperatives. This has significant implications for community engagement as the regulatory framework in the country has to balance the realisation of the rights of communities against the economic dependence on mining for development. In this vacuum, there has been a general disregard for the provisions of mining legislation by mining companies. This contested nature of mining versus community interests comes to the fore in the widespread battles between CSOs and mining communities.

A second, and related, insight relates to the South African government's sense of justice and fair play, as well as the will to maximise the distribution of mining benefits to the population at large. Generally, efforts made for broad-based empowerment initiatives for mine-affected communities by the Ministry of Mineral Resources in the country should be welcomed. However, the questionable value systems embedded within the ruling party have created doubt among mining communities that believe that the MPRDA has not led to the implementation and enforcement of compliance with the objectives of community engagement in the granting of licences to operate. This inaction by the Ministry of Mineral Resources has led to the escalation of conflict between civil society and mining companies.

Third, CSO mobilisation for the rights of mining communities points to a new era of community organising in South Africa in which activists have begun to challenge unilateral decisions taken by the government and mining companies. The struggles of mining communities differ from case to case. In most instances, some community mobilisation failed to gain momentum and a few – in relatively remote areas of South Africa – have garnered national attention for their sustained opposition to mining projects. Mobilisation by the Xolobeni and Lesetlheng communities provides good cases of the daily struggles of mining communities across South Africa that face exclusion. The October and November 2018 legal victories of the two communities over rural people's land rights emphasised that there must be consultation and consent before the land rights of citizens can be changed. Precarious mining communities managed to shift the balance of power in respect of court challenges with the legal support non-state institutions and community networks, such as MACAU, WAMAU, and MEJCON-SA. Thus, without the support of national CSOs and active support from various social formations, mining communities in South Africa will find it difficult to mobilise and succeed in the face of government and mining-company exclusion.

The dynamics between the business-centric goals of mining companies and community interests are involved. Negotiating a balance between the two will be necessary for future stability in mining communities. It is still early to determine to what extent civil society mobilisation will lead to long-term material gains for mining communities. Ultimately, fundamental change in the hostile relations between communities and mining companies in South Africa will depend on whether the government starts to support community development initiatives that have the blessing of communities. However, the controversial Traditional and Khoi-San Leadership Bill – seen by experts as an attempt to pre-empt the legal victories of mining communities – and the general attitude of the Ministry of Mineral Resources towards community land rights point out that passionate community mobilising against is long from being over. In the next chapters, I will bring local cases into a more general discussion.

References

Abuya, W.O. (2016). Mining conflicts and corporate social responsibility: Titanium mining in Kwale, Kenya. *The Extractive Industries and Society*, 3(2016): 485–493.

Acca (2018). Alternative Mining Indaba 2018. Available at www.accuhumanrights.org (accessed 17 August 2018).

Africa News (2018). Nigeria's Bodo community claims win over Shell after latest UK court ruling [online]. Available at www.africanews.com/2018/05/25/nigeria-s-bodo-community-claims-win-over-shell-after-latest-uk-court-ruling/ (accessed 19 June 2019).

African National Congress (ANC) (1977). *ANC speaks: Documents and statements of the African National Congress [1955–1976]*. London: ANC.

African National Congress (ANC) (2015). January 8th statement of the National Executive Committee of the ANC [online]. Available at www.anc.org.za/ancdocs (accessed 12 May 2019).

Anglo American (2019). Mining makes a difference [online]. Available at https://southafrica.angloamerican.com/~/media/Files/A/Anglo-American-South-Africa-V2/Attachments/media/mining-lekgotla-brochure/corporate-social-investment-mining-lekgotla-brochure.pdf (accessed 12 May 2019).

Attwood, B. (2003). *Rights of aborigines*. Crows Nest: Allen & Unwin.

Banks, G., Kuir-Ayius, D., Kombako, D. & Sagir, B.F. (2017). Dissecting corporate community development in the large-scale Melanesian mining sector. In: C. Filer & P. Le Meur (eds). *Large-scale mines and local-level politics between New Caledonia and Papua New Guinea* (pp. 217–228). Canberra: ANU Press.

Baxter, R. (2017). Mining Charter draft concerns – Chamber of mines. *Inside Mining*, 10(1): 32–33.

Becker, C. (2018). Amendments to Zimbabwe's indigenisation laws to open economy to foreign investment [online]. Available at www.mondaq.com/southafrica/x/691594/Inward+Foreign+Investment/Amendments+To+Zimbabwes+Indigenisation+Laws+To+Open+Economy+To+Foreign+Investment (accessed 19 June 2019).

Bench Marks Foundation (2016). Proposed changes to mining [online]. Available at www.bench-marks.org.za/press/proposed_changes_to_mining_charter.pdf (accessed 19 June 2019).

Berry, S. (2017). Struggles over land and authority in Africa. *African Studies Review*, 60 (3):105–125.

Bjuremalm, H., Gibaja, A.F. & Molleda, J.V. (2014). *Democratic accountability in service delivery: A practical guide to identify improvements through assessment*. Stockholm: International IDEA.

Black Economic Empowerment Commission (2001). *Black Economic Empowerment Commission Report*. Johannesburg: Skotaville Press.

Booi, Z. (2018). Dispossession without compensation the legacy for poor communities. *Daily Maverick* [online]. Available at www.dailymaverick.co.za/article/2018-06-11-dispossession-without-compensation-the-legacy-for-poor-rural-communities/ (accessed 19 June 2019).

Broome, R. (2010). *Aboriginal Australians: A history since 1788*. Crows Nest: Allen & Unwin.

Bruce, L. (2017). *Victory for mining community networks*. Centre for Applied Legal Studies, University of Witwatersrand [online]. Available at www.wits.ac.za/news/sources/cals-news/2017/victory-for-mining-community-networks.html (accessed 19 June 2019).

Cabe, M. (2018). Legal victories for communities [online]. Available at www.newframe.com/legal-victory-mining-communities/ (accessed 12 December 2018).

Cawood, F.T. & Minnitt, R.C.A. (1998). A historical perspective on the economics of the ownership of mineral rights ownership. *The Journal of the South African Institute of Mining and Metallurgy*, 98(November/December): 369–376.

Centre for Environmental Rights (2015). Joint media statement by civil society coalition on the MPRDA: Minister Ramatlhodi chooses a dangerous path [online]. Available at https://cer.org.za/news/joint-media-statement-by-civil-society-coalition-on-the-mprda-minister-ramatlhodi-chooses-a-dangerous-path (accessed 19 June 2019).

Chimhowu, A. (2019). The 'new' African customary land tenure. Characteristic, features and policy implications of a new paradigm. *Land Use Policy*, 81(February): 897–903.

Claassens, A. (2019). Xolobeni community being coerced into giving up land rights. www.customcontested.co.za/xolobeni-community-being-coerced-into-giving-up-land-rights/ (accessed 9 May 2019).

Comaroff, J.L. & Comaroff, J. (2018). Chiefs, capital and state in contemporary Africa. In: J.L. Comaroff & J. Comaroff (eds). *Politics of custom*. Chicago, IL and London: University of Chicago Press.

Dalaibuyan, B. (2015). *Mining, "social license" and local-level agreements in Mongolia*. Conference paper: International Conference on Perspectives on the Development of Energy and Mineral Resources Hawaii, Mongolia and Germany, University of Hawaii, Manoa. Available at www.researchgate.net/publication/303062239_Mining_social_license_and_local_level_agreements_in_Mongolia (accessed 11 May 2017).

Dalaibuyan, B. (2017). Local level agreements in Mongolia's resource sector. Lessons learned and the way forward. Available at https://resourcegovernance.org/sites/default/files/documents/nrgi-mongolia-agreement-briefing-english.pdf (accessed 19 June 2019).

Dale, M.O. (1997). South Africa: Development of a new mineral policy. *Resources Policy*, 23(1–2): 15–25.

Dauvergne, P. & LeBaron, G. (2014). *Protest Inc.: The corporatization of activism*. Cambridge: Polity Press.

Deloitte (2018). Tracking the trends 2018: The top 10 issues shaping mining in the year ahead. Available at www.mining.com/wp-content/uploads/2018/01/Deloitte-Tracking-the-Trends-Global-Mining-Study-FINAL.pdf (accessed 30 February 2019).

Department of Land Affairs (1997). *White paper on South African land policy*. Pretoria: Department of Land Affairs.

DMR (Department of Mineral Resources) (2002). Pretoria: DMR [online]. Available at www.dmr.gov.za/Portals/0/mineraland_petroleum_resources_development_actmprda.pdf (accessed 19 June 2016).

Economic Commission for Africa (2011). *Minerals and Africa's development: International study group report on Africa's mineral regimes*. Ethiopia: Economic Commission for Africa.

Evans, S. (2018). Landmark ConCourt judgment says mining rights do not trump lawful land occupier rights. *News24* [online]. Available at www.news24.com/SouthAfrica/News/landmark-concourt-judgment-says-mining-rights-do-not-trump-lawful-land-occupier-rights-20181025 (accessed 19 June 2019).

EITI (The Extractive Industries Transparency Initiative). 2019. Who we are [online]. Available at https://eiti.org/who-we-are#aim-of-the-eiti (accessed 6 June 2019).

Faku, D. (2016). Civil society takes mining concerns to UN. *Business Report*, 12 October [online]. Available at www.iol.co.za/business-report/economy/civil-society-takes-mining-concerns-to-un-2078801 (accessed 19 June 2019).

Fanon, F. (1963). *The wretched of the earth*. New York: Grove Press.

Ferguson, J. (2018). The image of mining: Changing public, employee, and customer perceptions. In: *Deloitte tracking the trends 2018: The top 10 issues shaping mining in the year ahead* [online]. Available at www.mining.com/wp-content/uploads/2018/01/Deloitte-Tracking-the-Trends-Global-Mining-Study-FINAL.pdf (accessed 28 April 2019)

Friedman, S. (2006). *Participatory governance and citizen action in post-apartheid South Africa*. Discussion Paper Series, No. 164. Geneva: International Institute for Labour Studies.

Gathii, J. & Odumosu-Ayanu, I.T. (2015). The turn to contractual responsibility in the global extractive industry. *Business and Human Rights Journal*, 1(1): 69–94.

Gilberthorpe, E. (2015). Silver Bullets or White Elephants? An assessment of the effectiveness of Community Share Ownership Trusts in Zimbabwe's Mining Industry. Dissertation, University of East Anglia. Available at www.eastanglia.academia.edu (accessed 12 August 2018).

Gill, M. (2017). Risk – why do people do what they do? *AUSIMM Bulletin*, April. Available at www.ausimmbulletin.com/feature/risk-why-do-people-do-what-they-do/ (accessed 12 March 2018).

Goldthau, A. (2018). *The politics of Shale Gas in Eastern Europe: Energy security, contested technologies and the social licence to frack*. Cambridge: Cambridge University Press.

Gottschalk, K. (2015). The legacy of South Africa's Freedom Charter 60 years later. *The Conversation*, 25 June [online]. Available at https://theconversation.com/the-legacy-of-south-africas-freedom-charter-60-years-later-43647 (accessed 19 June 2019).

Grant, J.A., Panagos, D., Hughes, M. & Mitchell, M.I. (2014). A historical institutionalist understanding of participatory governance and aboriginal peoples: The case of policy change in Ontario's mining sector. DOI:10.1111/ssqu.12115.

Groenewald, Y. (2018). Chamber of mines welcomes new minister Mantashe as 'man of integrity' [online]. Available at www.fin24.com/Companies/Mining/chamber-of-mines-welcomes-new-minister-mantashe-as-man-of-integrity-20180227 (accessed 19 June 2019).

Gufstafsson, M. (2018). *Private politics and peasant mobilization mining in Peru*. Cham: Palgrave MacMillan.

Gupta, D. (2017). *Protest politics today*. Cambridge: Polity.

Hall, R. (2010). Reconciling the past, present and future: The parameters and practices of *land* restitution in South Africa. In: C. Walker, A. Bohlin, R. Hall & T. Kepe (eds). *Land, memory, reconstruction and justice: Perspectives on land claims in South Africa* (pp. 17–40). Scottsville, VA: University of Kwazulu-Natal Press.

Harrison, P. (2016). The revolt of South Africa's metropoles: A revolution of rising expectations. *The Conversation*, 5 September [online]. Available at https://theconversation.com/the-revolt-of-south-africas-metropoles-a-revolution-of-rising-expectations-64617 (accessed 14 August 2018).

Harvey, B. (2018). The power of local level agreements. *AusiIMM Bulletin* [online]. Available at www.ausimmbulletin.com/feature/power-local-level-agreements/ (accessed 19 June 2019).

Haslam, P.A. (2018). Beyond voluntary: State–firm bargaining over corporate social responsibilities in mining. *Review of International Political Economy*, 25(3): 418–440.

Hayman, G. (2014). Reversing the curse: The global campaign to follow the money paid by oil and mining global industries. In: G. Sweeney, R. Dobson, K. Despota & D. Zinnbauer (eds). *Corruption report: Climate change*. London: Earthscan.

Hepburn, S. (2015). *Mining and energy law*. Cambridge: Cambridge University Press.

Hill, N. & Maroun, W. (2015). Assessing the potential impact of the Marikana incident on South African mining companies: An event method study. *South African Journal of Economic Management Sciences*, 18(4): 586–607.

Huni, S. (2017). *An analysis of the mining regulatory framework and its impact on the drive for beneficiation in Zimbabwe between 2009 and 2016*. Masters dissertation. UP [online]. Available at https://repository.up.ac.za/bitstream/handle/2263/65657/Huni_Analysis_2018.pdf?sequence=1 (accessed 19 June 2019).

Hunter, T. (2015). *Regulation of the upstream petroleum sector: A comparative study of licensing and concession systems* (New Horizons in Environmental and Energy Law Series). Cheltenham: Edward Elgar Publishers.

Jackson, M., Watson, K., Watson, S., Goodison, J. & Padoa, J. (2015). South Africa: Broad-based black economic empowerment: Then and now – Circular 1 [online]. Available at www.mondaq.com/southafrica/x/372086/Corporate+Commercial+Law/BroadBased+Black+Economic+Empowerment+Then+And+Now+Circular+1 (accessed 19 June 2019).

Johnson, R.W. (2015). *How long will South Africa survive? The looming crisis*. Johannesburg and Cape Town: Jonathan Ball.

Kane-Berman, J. (2017). *Mining in SA: Then, now, and into the future*. Institute of race Relations [online]. Available at www.politicsweb.co.za/documents/mining-in-sa-then-now-and-into-the-future--irr (accessed 5 January 2019).

Kapdi, N, Sulaiman, S., Twala, Z. & Mthi, M. (2015). Wholesale licensing in the downstream petroleum industry in South Africa and certain other African jurisdictions [online]. Available at www.mondaq.com/southafrica/x/438780/Oil+Gas+Electricity/Wholesale+Licensing+In+The+Downstream+Petroleum+Industry+In+South+Africa+And+Certain+Other+African+Jurisdictions (accessed 19 June 2019).

Kiewit, L. (2019). Contentious traditional leadership Bill passed [online]. Available at https://mg.co.za/article/2019-01-11-00-contentious-traditional-leadership-bill-passed (accessed 19 June 2019).

Lahiff, E. (2007). Willing buyer, willing seller: South Africa's failed experiment in market-led agrarian reform. *Third World Quarterly*, 28(8): 1577–1597.

Lang, A. (1999). *Separate development and the Department of Bantu Administration in South Africa*. Work from the Institute of African Studies No. 103. Hamburg: Composite Foundation German Overseas Institute.

Leon, P., Cannon, A., Maxwell, I. & Ambrose, H. (2018). Upheaval and uncertainty in mineral regulation in parts of Africa resurgence of resource nationalism highlights the importance of investment treaty protections [online]. Available at https://hsfnotes.com/arbitration/2018/05/11/upheaval-and-uncertainty-in-mineral-regulation-in-parts-of-africa-resurgence-of-resource-nationalism-highlights-the-importance-of-investment-treaty-protections/?utm_source=Mondaq&utm_medium=syndication&utmcampaign=LinkedIn-integration (accessed 12 December 2018).

Mabe, M. (2019). Xolobeni residents divided over mine. *Sowetan*, 17 January, p. 5.

Mabuza, E. (2018). 'Massive victory' for small community in mining rights case – LHR [online]. Available at www.timeslive.co.za/news/south-africa/2018-10-25-massive-victory-for-small-community-in-mining-rights-case-lhr/ (accessed 12 December 2019).

Mandela Institute. (2017). *Public regulation and corporate practices in the extractive industry: A South-South advocacy report on community engagement*. Johannesburg: University of Witwatersrand.

Matebesi S.Z. & Botes, L.J. (2017). Party identification and service delivery protests in the Eastern Cape and Northern Cape, South Africa. *African Sociological Review*, 21(2):81–99.

Matebesi, S.Z. & Marais L. (2018). Social licensing and mining in South Africa: Reflections from community protests at a mining site. *Resources Policy*, 59(December): 371–378.

Matyszak, D & Research and Advocacy Unit. (2013). *Digging up the truth': The legal and political realities of the ZIMPLATS*. Harare: Research and Advocacy Unit.

Miller, D.L.C. & Pope, A. (2000). *Land title in South Africa*. Cape Town: Juta & Co.

Mitchley, A. (2018). High court rules in favour of Xolobeni community in historic mining rights case. Mail and Guardian. Available at https://mg.co.za/article/2018-11 (accessed 12 December 2018).

Mitchell, A., Moalusi, L., van der Want, M., Bryson, S., Picas, C. & Verwey, J. (2012). The Avatar syndrome: Mining and communities. *The Southern African Institute of Mining and Metallurgy*, 112(February): 151–155.

Moalusi, L. & Malesa, G. (2019). Looking ahead: Policy and regulation framework for South Africa's mining industry [online]. Available at www.businesslive.co.za/bd/national/2019-01-28-looking-ahead-policy-and-regulatory-framework-for-sas-mining-industry/ (accessed 12 March 2019).

Modise, L.J. (2017). The notion of participatory democracy in relation to local ward committees: The distribution of power. *In die Skriflig*, 51(1): a2248. DOI:10.4102/ids.

Moosa, F. (2018). The Xolobeni community is fighting for a voice in mining decisions [online]. Available at www.thedailyvox.co.za/the-xolobeni-community-is-fighting-for-a-voice-in-mining-decisions-fatima-moosa/ (accessed 7 December 2018).

Mostert, H., Chisanga, M., Howard, J., Mandhu, F., Vander Berg, M. & Young, C. (2016). Corporate social responsibility in the mining industry in Namibia, South Africa, and Zambia. In: L.K. Barrera-Hernández., B. Barton, L. Godden, A. Lucas & A. Rønne (eds). *Sharing the costs and benefits of energy and resource activity – Legal change and impact on communities* (pp. 93–112). Oxford: Oxford University Press.

Murombo, T. (2013). Regulating Mining in South Africa and Zimbabwe: Communities,the environment and perpetual exploitation. *Law, Environment and Development Journal*, 9(1): 31.

Nelwamondo, T. (2016). History of success and challenges of SA: Community Trust [online]. Available at https://sawea.org.za/wp-content/uploads/2016/05/Community-Trust-in-South-Africa-SAWEA-SAPVIA-23-MAY-2016.pdf (accessed 2 August 2017).

Nicolson, G. (2015). Chamber of mines AGM: A necessary friction. *Daily Maverick*, 21 May [online]. Available at www.dailymaverick.co.za/article/2015-05-21-chamber-of-mines-agm-a-necessary-friction/ (accessed 19 June 2019).

O'Faircheallaigh, C. (2012). Community development agreements in the mining industry: An emerging global phenomenon. *Community Development*, 44(2): 222–238.

Oxfam (2015). Community consent index 2015. Oil, gas, and mining company public positions on Free, Prior, and Informed Consent. 207 Oxfam Briefing Paper. Available at www-cdn.oxfam.org (accessed 24 May 2019).

Pather, R. (2018). It's not just Xolobeni: What the Australian mining company did in the Western Cape. *Mail & Guardian*, 24 April [online]. Available at https://mg.co.za/article/2018-04-24-australian-mining-company-already-faces-resentment-in-western-cape (accessed 19 June 2019).

Pegg, S. & Zabbey, N. (2013). Oil and water: The Bodo spills and the destruction of traditional livelihood structures in the Niger Delta. *Community Development Journal*, 48(3): 391–405.

Petras, J.F. & Veltmeyer. H. (2019). Neoliberalism and social movement in Latin America: Mobilizing the resistance. In: B. Berberoglu (ed.). *The Palgrave handbook of social movements, revolution, and social transformation* (pp. 177–211). Cham: Palgrave McMillan.

Ramatji, K.N. (2013). A legal analysis of the Mineral and Petroleum Resources Development Act (MPRDA) 28 of 2002 and its impact in the mining operations in the Limpopo

Province [online]. Available at http://ulspace.ul.ac.za/bitstream/handle/10386/979/ramatji_kn_2013.pdf?sequence=1 (accessed 15 July 2018).

Rametse, M.S. (2016). *The significance of the Freedom Charter in the ideological debates within the ruling ANC Alliance in South Africa*. African Studies Association of Australasia and the Pacific (AFSAAP) Proceedings of the 38th AFSAAP Conference: 21st Century Tensions and Transformation in Africa, Deakin University, 28th–30th October 2015 [online]. Available at http://afsaap.org.au/assets/Mochekoe_Stephen_Rametse_AFSAAP2015.pdf (accessed 19 June 2019).

Ramphele, M. (2019). Xolobeni and the iron history [online]. Available at www.news24.com/Columnists/Mamphela_Ramphele/xolobeni-and-the-irony-of-history-20190305 (accessed 19 June 2019).

Reuters. (2018). Land, mining uncertainty threatens South Africa growth – Moody's [online]. Available at www.iol.co.za/business-report/economy/land-mining-uncertainty-threatens-south-africa-growth-moodys-15692605 (accessed 7 September 2018).

Riofrancos, T (2019). What comes after extractivism? *Dissent*, 66(1): 55–61.

Roeder, R.W. (2016). *Foreign mining investment law: The cases of Australia, South Africa and Colombia*. Cham: Springer.

Rutledge, C. (2015). Why mining communities will take government to court. *Groundup*, 21 May [online]. Available at www.groundup.org.za/article/why-mining-communities-will-take-government-court_2961/ (accessed 19 September 2018).

Sander, I. (2018). The future of work. In: *Deloitte. Tracking the trends 2018: The top 10 issues shaping mining in the year ahead* [online]. Available at www.mining.com/wp-content/uploads/2018/01/Deloitte-Tracking-the-Trends-Global-Mining-Study-FINAL.pdf (accessed 19 September 2018).

Short, D. (2016). *Reconciliation and colonial power: Indigenous rights in Australia*. New York: Routledge.

Sidley, K. (2015). Community trusts the best model to exploit resources. *Business Day*, (24 June 2015).

SIOC (Sishen Iron Ore Company) (2019). Who we are [online]. Available at www.sioc-cdt.co.za/about/who-we-are/ (accessed 19 June 2019).

Stoltenborg, D. & Boelens, R. (2016). Disputes over land and water rights in gold mining: Case of Cerro de San Pedro, Mexico. *Water International*, 41(3): 447–467.

Suttner, R. & Cronin, J. (1986). 30 years of the Freedom Charter. *Transformation*, 6: 73–86.

Terwindt, C. & Schliemann, C. (2017). *Tricky business: Space for civil society in natural resource struggles. Publication Series on Democracy*. Berlin: Heinrich Böll Foundation or the European Center for Constitutional and Human Rights [online]. Available at www.boell.de/sites/default/files/tricky-business.pdf (accessed 19 September 2018).

Truter, J. (2011). South Africa: Mining rehabilitation – A regulated activity [online]. Available at www.mondaq.com/southafrica/x/155372/Mining/Mining+Rehabilitation+A+Regulated+Activity (accessed 19 September 2018).

Van der Merwe, C. (2011). Shabangu withdraws Xolobeni mining licence. *Engineering News* [online]. Available at www.engineeringnews.co.za/print-version/shabangu-withdraws-xolobeni-mining-licence-2011-06-07 (accessed 14 November 2018).

Van der Schyff, E. (2012). South African mineral law: A historical overview of the State's regulatory power regarding the exploitation of minerals. *New Contree*, 42(July): 131–153.

Vivoda, V. & Kemp, D. (2017). *Comparative analysis of legal and regulatory frameworks for resettlement in the global mining sector*. Brisbane: Centre for Social Responsibility in Mining (CSRM), The University of Queensland.

4 Royal Bafokeng Nation

A model of community-based natural-resource management

Introduction

The Royal Bafokeng Nation (RBN) and the Royal Bafokeng Nation Development Trust (RBNDT) are hailed as a model for mineral resource control at the community level, an area that has gained much recognition in the international literature. As a result, a number of studies have examined RBN by focusing on family law and the law of succession (Coertze, 1990), the possible role of traditional leadership in development planning (Nthau, 2002), legal battles (Manson & Mbenga, 2003; Cook, 2013), poverty alleviation strategies (Modipa, 2007), corporatisation of a tribal authority (Cook, 2009), platinum wealth and community participation (Mnwana & Akpan, 2009), the tensions and contradictions inherent in the RNB's status as both community and corporation (Cook, 2011), the community control of mineral wealth (Mnwana, 2012), the role of education (Gaborone, 2014), resource curse (Thompson, 2015), and satisfaction with services provided by the Royal Bafokeng Administration (RBA) (Flomenhoft, 2019).

Some of the earlier contributions claimed that little was known about the real character of the community control of natural resources in their study communities (Bench Marks Foundation 2007; Slack, 2009). However, over time the literature has become replete with cases that show the difficulty for mineral-rich indigenous societies to manage and distribute royalties (O'Faircheallaigh, 1998). Others have argued that community-based natural-resource management is susceptible to elite capture (by traditional leaders, politicians, and local leaders) (Platteau & Gaspart, 2003; Lucas, 2016) or patronage and collusive networks (Mrema, 2017). Williams and Le Billion (2017) have also revisited the challenges of resource curse by conducting nuanced and policy-relevant case studies analysing patterns of corruption around natural resources.

In a recent report of Corruption Watch (2019), which details why the royalties and benefit-sharing system appear to be failing, it is emphasised that mining has generated many jobs and contributed significantly to industrial development in South Africa. However, it has also "contributed to corruption, environmental harm and labour exploitation, while communities living on resource-rich land have been deprived of the benefits due to them" (Corruption Watch, 2019: 1).

This chapter examines, empirically, how the RNB reconciles customary institutions with democratic governance institutions to maintain a social licence. Here, the emphasis is on the relationship between the policy level and the decision-making levels in path dependency, an under-researched component of path dependency (Matebesi & Marais, 2018). I argue that the abundance of platinum resources cannot alone explain the success of RNB in the management of mining royalties. I insist that this is a misleading narrative, as resource abundance does not necessarily equate the prudent management of royalties to the benefit of locals. It is evident that RNB became locked in the legacies of respecting the linkages between customary governance institutions and modern institutions of governance as they evolve.

Considerable attention is also given to the perspective of local community members in Phokeng (the capital of RNB) to examine their views on the local benefits of the royalties managed by RBN. In this regard, questions posed by Cook (2011: S156) about the RBN management of mineral royalties are significant:

> If the benefits are collective, how is the effect on individuals measured? What, specifically, constitutes evidence of service delivery and equitable distribution of communal resources? Is it the annual budget and spending priorities of the nation? Is it anecdotal feedback from the community members as they represent themselves to kgotha-kgothe and increasingly to the media?

Demographic and socio-economic context of RNB

RBN is a group of over 100,000 Setswana-speaking people, living on their ancestral land in 29 villages and peri-urban areas in Rustenburg, North West Province of South Africa. Geographically, RNB falls under the Rustenburg Local Municipality, which forms part of the greater Bojanala Municipality. The population of Rustenburg Local Municipality was 645,000 in 2017, which constituted 1.1% of the then total population of South Africa. The municipality had significantly more males (54.2%) relative to South Africa (48.9%). The largest share of the Rustenburg population is within the younger working category of 25–44 (41%). In 2017, the unemployment rate stood at 25.1%, while 13.7% of the households were earning R30,000 per annum. In respect of education, 31.2% had no formal schooling. In respect of poverty trends in Rustenburg, 42% of the population lived in poverty in 2017, compared to 52.8% in 2007. The municipality seems to be doing well in respect of services delivery, as only 0.9% households had no formal piped water supply and only 10.1% were responsible for their own refuse removal (Rustenburg Local Municipality, 2019) (Figure 4.1).

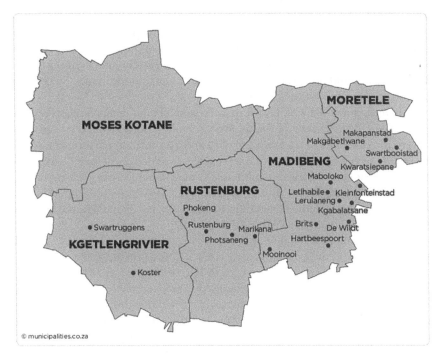

Figure 4.1 Map of Rustenburg and surrounding areas.

The institution of traditional leadership in South Africa

There has been growing criticism of the role and relevance of traditional authorities in a modern, capitalist, globalised world. However, some scholars note that while the formation of a modern state may have restricted traditional leaders' discretion as sovereigns, "traditional authority is not overwritten but rather refined, transformed and stabilised in the process of state formation" (Tieleman & Uitermark, 2018: 1). Others regard them as "agents of the regime" (De Kadt & Larreguy, 2018: 382).

Notwithstanding this criticism, the South African government had made major strides over the past few years to transform the institution of traditional leadership. For example, in a briefing to the Parliamentary Portfolio Committee on Cooperative Governance and Traditional Affairs (COGTA) in November 2014, the Department of COGTA instructed the Committee that the institution of traditional leadership was recognised in Section 212 of the Constitution of South Africa. Furthermore, the White Paper on Traditional Leadership and Governance and the Traditional Leadership and Governance Framework Act of 2003 aimed to transform the institution of traditional leadership. The White Paper required that provision be made for the allocation of additional roles to traditional leadership. Members felt that municipalities were undermining traditional leaders and traditional councils. Municipalities did not take traditional councils seriously, and their representation on municipal councils was merely ceremonial. It was evident to members that the issue of the roles and functions of traditional leaders was a work in progress. It was generally submitted that traditional leaders could play key roles on issues, including, for example, job creation and housing provision (PMG, 2014).

The Department of COGTA felt that the issue of land deserves attention. For instance, the development of South African land arrangements removed most powers and capacities of traditional leaders and gave it to the government. Later, the powers were returned to traditional leaders as added capacities. Contrary to the customary law of African communities, traditional leaders were managing land on behalf of the South African government. As a result, traditional leaders played a central role in the democratisation process, particularly in the development of rural communities (PMG, 2014).

Fast forwarding to 2019, it seems that the role of traditional leaders in South Africa is not yet clear, as each province has its own provincial law relating to traditional leadership and governance (PMG, 2014). The Traditional Leadership and Governance Framework Act of 2003 was developed to provide a national framework for the role of traditional leadership within the new system of democratic governance (Corruption Watch, 2019). This demonstrates that the institution of traditional leadership in South Africa has undergone "fundamental legislative changes throughout the different historical phases of pre-colonial, colonialism, apartheid and democratic dispensation" (Khunou, 2011: 290).

Historical path formations of RBN

The history of RBN is fascinating in many respects. However, despite the long recognition that history plays an important role in development in general, many scholars – except Cawood and Minnitt (1998) and Cook (2005, 2011) – restated this history without explanations of how history influenced the process of RNB to become what is unofficially known as a model of community-based natural-resource management. In Table 4.1, I outline the history of RBN, but it is by no means a complete chronicle of the elaborate history of RBN, which has been presented excellently elsewhere (for example, Coertze, 1990; Cook, 2005, 2011, 2013; Mnwana & Akpan, 2009; Mbenga & Manson, 2010; Gaborone, 2014).

Two major features of the Bafokeng's quest to assert their land rights can be identified. First was the challenge of RBN to deal with the significant unjust system of colonialism, which forbade Africans from buying or owning land during the 1840s. As a result, the Bafokeng were legally excluded from owning their communal land (Cawood & Minnitt, 1998). This decision by the Afrikaans Voortrekkers established a legal development pathway for the Bafokeng. But how?

At the time, Kgosi Mokgatle, who ruled the Bafokeng between 1834 and 1891, "understood the need to regain some form of legal security of tenure" (Molotlegi, 2013). Subsequently, he arranged with white missionaries in the area to buy land on the Bafokeng's behalf, registering the farms in their names and holding them in trust for the community. Interestingly, Kgosi Mokgatle managed to invest the offerings he solicited from young Bafokeng who got jobs at the newly opened Kimberly diamond mine in the Northern Cape Province of South Africa. Thus, Kgosi Mokgatle was able to employ, in line with what proponents of the evolutionary governance theory (EGT) posit, an understanding of the context in which the intervention is to materialise (Beunen et al., 2016). According to Kgosi Molotlegi (2013), Kgosi Mokgatle used a traditional form of

Table 4.1 Selected timeline of historical events related to RBN

Year	Event
1867	• Establishment of the Hermansburg Mission established by Missionary Christoph Penzhorn who holds the land in trust for Kgosi Mokgatle. He also introduced education to the Bafokeng.
1877	• A strong regiment of 500 men go to work on the diamond fields of Kimberley and agree to deposit a share of their wages into a trust to buy land for RBN.
1913	• Land Act permits the blacks to own land, and so the tradition is executed under August Molotlegi.
1924	• Discovery of Merensky Reef which holds the world's largest supply of platinum-group metals
1960	• Platinum mining begins. • Bophuthatswana homeland created for Batswana set-up, presided over by Lucas Mangope.
1978	• Bafokeng receives first royalty payment, which could easily be manipulated to the mining giant's advantage.
1985	• "The Deeps" mining area opened with an increased royalty payment, which led to conflict between the Impala Mining Company and the Bophuthatswana government.
1988	• Coup staged to overthrow Mangope but failed. Kgosi Edward Lebone I Molotlegi forced into exile.
1990	• George Mokgwaro Molotlegi installed as ruler.
1994	• Kgosi Lebone I, his wife, and children return to Phokeng. Then begins the tug of war for better royalties between the mining giants and the Kgosi who acted on behalf of the Bafokeng
1999	• A settlement is finally reached, giving Bafokeng 22% taxable income on all operations which they use to bring to life Vision 2020 and Lebone, a school of excellence, is established
2004	• Legally sought to have their revenues exempted from nationalisation after the Minerals and Petroleum Resources Development Act (No. 28 of 2002) came into effect in May 2004
2010	• Lebone II College of the Royal Bafokeng campus is established on Tshufi Hill. • The Communal Land Rights Act (No. 11 of 2004), which sought to shift control of communally administered land from tribal councils to government-controlled land rights boards and was contested by RBC from the outset, is declared unconstitutional.

Source: Adapted from Cawood and Minnitt (1998), Cook (2005, 2011), Mbenga and Manson (2010), Molotlegi (2013), and Gaborone (2014).

social organisation and the youth regiments to handle a particularly modern economic problem. Thus, the ability of Bafokeng leaders to handle complex contemporary challenges – and exceed expectations – has characterised the Bafokeng from that time to the present day.

A second major, and more fundamentally, feature of the Bafokeng history is the many legal battles it fought with the apartheid government, as well as the democratically elected government of 1994 (Cawood & Minnitt, 1998; Gaborone, 2014). For example, Cook (2011) suggests that the history of land acquisition by the Bafokeng highlights how RBN established itself as a private corporate landowner as early as the late 19th century. By the time platinum was discovered in Bafokeng territory in 1924 (Mbenga & Manson, 2010), RBN was able to lease parts of their land to various mining companies. After the creation of the homeland system in South Africa, another threat was looming to the land rights of the Bafokeng.

This time the threat came in the form of the president of Bophuthatswana, Lucas Mangope, who frantically tried to negotiate contracts directly with the mining company Impala Platinum on behalf of the Bafokeng. This led to a two-pronged battle between the Bafokeng and Bophuthatswana, as well as the Bafokeng and Impala Platinum. After an out-of-court settlement, which was hailed as a landmark decision against major mining in South Africa, the Bafokeng got recognition for using a legal recourse to their advantage (Cook, 2011; Gaborone, 2014).

Interestingly, this shows how a tribal-governance system has managed to force the government and the mining industry through the litigation to the bargaining table, to define the boundaries of discussions, and to come out victorious. However, the litany of threats to the control of Bafokeng's land and its resources persisted, with the advent of the democratic government in 1994. The two pieces of legislation that particularly threatened the status of RBN were the Minerals and Petroleum Resources Development Act (No. 28 of 2002) and the Communal Land Rights Act (No. 11 of 2004). The former sought to change the legal ownership rights of minerals in South Africa, while the latter had the intention of shifting control of communally administered land from tribal councils to the land rights boards under the control of the government (Cook, 2011). The Communal Land Rights Act was declared unconstitutional in 2010, after four rural communities, including RBN, challenged its constitutionality, arguing that it would undermine their right to tenure security, as set out in the South African Constitution (Hull, 2018).

It is evident that despite the legal recognition afforded to tribal independence in the past and during post-apartheid South Africa, the conduct of those in power at any given time has often been directed towards the destruction of these tribal rights. Accordingly, for successive generations, the RBN established and maintained the legal approach as a dominant approach to the threats to its land rights and mineral rights. This approach of the RNB leaders exhibits path dependence. This lock-in in the legal approach has been an effective strategy for RBN, which, by 2000, was receiving 22% in mining royalties of all operations on the lands

RNB owned. This, Cook (2005) believes, reinforced the process of the corporatisation of RBN (Cook, 2005) that occurred over many decades. I now turn to the governance of RBN.

Governance and engagement via path-dependent institutions

In this section, I briefly illuminate the governance institutions underpinning the tribal acumen of the Bafokeng and the challenges of persistent adaptation to the evolutionary reforms of modernity. The shift from a predominantly agricultural to mining community did not alter the governance approach of RBN. The goal has always been to sustain the salient tribal structures that articulate the vision of RNB. These structures of the traditionally governed Tswana-speaking community of RBN include a hereditary king and 72 hereditary *dikgosana* (ward headmen) (Mnwana & Akpan, 2009; Cook, 2011; Gaborone, 2014).

In what represents the broadening of its pathways of community engagement, the RNB is also governed by elected structures, including Bafokeng councillors representing the five regions of the RBN (Manson & Mbenga, 2010). The trend towards an increasingly corporatist pathway of governance is witnessed in the *Kgothakgothe*; the highest decision-making structure remains the general meeting of Bafokeng citizens biannually (Comaroff & Comaroff, 2009). According to Kgosi Molotlegi (2013), all major decisions concerning land and the strategic assets of RBN must be agreed to by resolution at the Kgothakgothe. Furthermore, seemingly, this general meeting has the right to reject proposals by the Administration, the Supreme Council of Dikgosana and councillors, and even the king (Molotlegi, 2013).

There is no doubt that such an extensive network of engagement has the potential to support an inclusive vision that fosters trust and strengthens the relationship between the Bafokeng tribal leadership and citizens. However, the tribal consultative system has also been extended by professional administrative structures that are driven by long-term planning processes and informed by evidence-based empirical research.

Like the RBA, the Royal Bafokeng Institute facilitates the delivery of quality education and training, and the Royal Bafokeng Enterprise Development and the Royal Bafokeng Holdings (RBH) manage the commercial assets of the RBN spread across mining, financial services; infrastructure, oil and gas services; and industrial sectors. The shares accrued from the different entities are held by the RBNDT, which is also responsible for social delivery entities (Comaroff & Comaroff, 2009; Cook, 2011; Molotlegi, 2013).

Research suggests that path dependence presents both challenges and opportunities, including the possibility that efforts can generate self-perpetuating outcomes (Tekwa et al., 2019). RBN has demonstrated that path dependence has provided an opportunity to develop service-delivery partnerships with the Rustenburg Local Municipality. This is done by directing the funds from various investment portfolios to provide services like the maintenance of physical assets and the rendering of municipal services (Gaborone, 2014).

Thus, the pathway of reconciling traditional and modern institutions of governance, I argue, "locked-in" RNB in the practice of systematically engaging long-term planning and, thereby, protecting the very foundation of what led to be described as a model of community-based natural-resource management. I further argue that the failure of the current institutional arrangements to develop the underlying conditions of trust that ensure improved economic conditions of residents may lock RNB into a specific innovation trajectory.

In sum, it appears that over several decades, RNB maintained an SLO through an elaborate consultation process, helping the Bafokeng citizens to feel connected to their Tswana customs, as well as embracing modern practices of governance. Similarly, RNB assets have grown (and so have the names from Barefoot Billionaires to the Richest Tribe in Africa). However, a recurring question when considering the RNB pathways of civic engagement is: to what extent do Bafokeng citizens have credible opportunities to influence decisions concerning the identification, leveraging, and mobilisation of community resources? This is a point which I return to later.

The case of RBN

In this section of the empirical findings, I look closely at the recurrent themes that arose from the fieldwork and survey residents in Phokeng. In so doing, the focus is on the in-depth interview with the Chief Executive Officer of RBNDT, Mr Obakeng Phethwe, and community members. As stated in Chapter 1, the survey seeks to understand the knowledge of residents about the community development trust (CDT), perceptions about community consultation and participation, local benefits, and protest participation, as well as the changes they would like to see in the future. A total of 200 respondents took part in the survey, with 56% ($n = 112$) of them being male and 44% ($n = 88$) female. It should be noted that the names "RBN" and "RBNDT" are used interchangeably.

The modus operandi of RBNDT/RBN

Several studies have shown the importance of enabling conditions in shaping an SLO, including a commitment to fair relations with communities and the public in negotiating local agreements or benefits (Lacey et al., 2017). In this regard, where enabling conditions to constrain the process of community acceptance, a need arises to disrupt that pattern. In the case of RBN, despite incremental changes in the legislative and policy environment in South Africa, the tribal authority leadership remained largely "locked-in" to past policies and actions aimed at reconciling traditional and modern institutions of governance.

This study discovered that the management of RBNDT – like all the other institutions of RBN – is premised on democratic principles of transparency and accountability. For example, the board of RBNDT includes both tribal and elected representatives of the community, who are responsible for strategic planning and the oversight of the CEO. The board has to ensure that the CEO carries out the plans of the board of trustees. The RBNDT seems to be proud of the historical trajectory of RBN, which serves as a template for dealing with the challenges of today. This is reflected by the CEO of RBNDT who stated that the history of the RBN has always centred on asset management. This started more than 100 years ago. Whatever you read about RBN, you will observe that the issue of land acquisition is a dominant theme. For us, we are grateful for the wisdom of our forefathers. As you will realise, there was no template for the early leaders of RNB to begin with, but they have set this nation on a path of economic emancipation, thereby achieving the Bafokeng aspiration of nurturing customary practices in a modern capitalist environment. In fact, each generation followed the same development trajectory of respecting the need to build sufficient investments for future generations. Bafokeng knew a decade ago about the significance of investment. But as we gradually corporatised our business affairs, we never forgot our traditional roots.

Granted, this statement from the CEO is wide open to interpretation. The results of the survey show that respondents considered that they have extensive knowledge ($n = 142$; 74.9%), little knowledge ($n = 43$; 21.3%), and no knowledge of RBNDT ($n = 15$; 7.8%). Those who had extensive knowledge and little knowledge ($n = 185$) were asked whether they are satisfied with the organisational structure of RBNDT; 67% ($n = 124$) indicated they are satisfied.

Similarly, the majority of the participants of the qualitative study indicated that they have extensive knowledge about RBNDT and are satisfied with its current organisational structure. As one community leader noted, "You cannot fix what is not broken. I reckon that the entire RBN organisational structure can compete with any private company in South Africa." Further responses to the question emphasised the "consistency," "innovativeness," and "vision" of RBN as standing out for them. A business owner went further and stated (in what has racial undertones), "after decades of being undermined as black people in this country, we can pride ourselves on the achievements of RBN."

When asked what makes RBNDT unique in comparison to other community trusts in South Africa, the CEO responded with great conviction that the emphasis of RBN had been consistently on investment and diversification, while other community trusts embark on the route of consumerism. He explained further that when mines began to give royalty shares to tribal authorities, most adopted several destructive tendencies. First, most of the royal family members at the time had no interest in finding a job after the sudden windfall from the mines. Second, the younger ones also started to drop out of school. Third was the fighting for a leadership position within the traditional authorities, with the sole aim of controlling the resources of the tribe. He further indicated that the forefathers of the Bafokeng were guided by the mantra of "Platinum Today, but what about Tomorrow." In what exemplifies path dependence, he emphasised how RNB spoke about "diversification when it was not fashionable," and this has become deeply ingrained in the *modus operandi* of the Bafokeng. As he reflects:

> We may be a tribal authority, but our governance approaches evolve, but not our culture. In fact, our history littered with elements of market processes and distinctive policy initiatives that have made us what we are today. For

any income we generate, our operational model is that at least 30% should be invested. It is not always possible, but it has proven to be a useful approach. We believe strongly that asset management provides a systemised approach to capital protection. We strive to have a better balance sheet all the time. Take, for example, in 2003, 95% of our total income came directly from mining, yet today [May 2018] marks six years without having received any dividends from mining. Our strategy is premised on the analogy of an ant. *Re tshwana le tsie. Ra itse gore mariga ke nako ya tlala* [We are like ants. We know that the hunger season is lying ahead]. That is why we do like taglines such as "Africa's richest tribe." We would rather be called an "organised" community. This is woven in our fabric.

The literature on path dependency conceives organisational acts as historically conditioned (Van Assche et al., 2014), while valuable insights into the dynamic nature of such eventually entrapping or lock-in processes are provided by contemporary evolutionary governance theorists (Van Assche et al., 2013; 2014; Beunen et al., 2016). What is perhaps important in this regard is to note that the source of the performative effects of governance – as illustrated by RBNDT in this study – is found in the traditional governance institutions of RBN. Increasingly, though, scholars have highlighted that the existence of traditional leaders in a democratic dispensation poses a challenge to the constitutional democracy in South Africa (Maseko, 2015).

In the following section, the focus turns to determining perceptions about the consultation and participation of citizens in the decision-making processes of the RBNDT. In particular, it seeks to determine the extent to which citizens in Phokeng are satisfied with the consultation and participation processes in line with the proliferation of consultation requirements in post-apartheid South Africa.

Consultation and citizen participation in Phokeng

EGT has been postulated "as useful for analysing existing forms of citizen participation in governance as well as proposals and the potential for new forms" (Van Assche et al., 2014: 89). However, "community engagement is a crucial component by which acceptance in the form of a social licence may be granted or not" (Measham & Zhang, 2019: 365). While the integrated path dependency model of Moffat and Zhang (2014) recognises the importance of consultation with local community members, the EGT helps us understand the reasons institutionalised forms of communication and consultation are critical to organisational success. As can be observed, the consultation and communication strategy of RBN is primarily based on the Bafokeng tribal institutional arrangements. Despite the wealth of information on RBN consultation structures, "the extent to which mineral-rich 'tribal' authorities are capable of engendering broad-based participation in the utilisation of mineral revenues remains less examined" (Mnwana, 2014: 827).

This section focuses on the procedural fairness of RBNDT, which, as stated in Chapter 2, was measured by two items that focus more on the collective than the individual respondent: "I believe RBNDT listen to and respect our opinions," and "I believe RBNDT is prepared to change its practices in response to community sentiment" (Moffat & Zhang, 2014, Measham & Zhang, 2019).

In concurring with the literature on the consultation processes of RBN, the CEO further elaborated on the importance of community engagement and the reason behind each consultative structure. He noted "Community engagement is not an easy process. However, it is imperative to understand and appreciate the needs of our residents fully."

For the CEO, the extensive and highly participatory consultation process has been effective in promoting community empowerment. The process, according to the CEO, begins with the *metse le metsana* meeting (village-based meetings), held each Sunday to listen to community concerns as well as provide feedback on developments within RNB. Then follow the regional meetings called *Dumela Phokeng* ("Hello Phokeng"), comprising three to four villages. The chief facilitates the *Dumela Phokeng* meetings, which take place annually between April and May to solicit general development ideas from the communities. The *Kgothatgothe* is held in November or December and is attended by an average of 3,000–5,000 individuals. A notable aspect of this *kgothagothe* meeting is that children are being entertained at the venue of the meeting through games and learning activities. The purpose of these activities for the children is *go ba ruta ka ga botlhokwa ba pitso ya kgotla* [to teach them about the significance of the Supreme Council meeting]. RBN also makes use of various online communication platforms, and print media, including the *Segoagoe* magazine, published quarterly (interview with CEO of RBNDT, 2 May 1998).

The extant literature on aspects of democracy, rights, participation, and accountability – often found lacking in traditional forms of governance – seems not to be the case with RBN. For example, Cook (2005) points out that the capacity of RBA to provide services, infrastructure, and university scholarships enhanced the image of RBN among Bafokeng villagers. Since liberal democratic structures and traditional forms of governance often exist in tension with one another (Cook, 2005), the standard interpretation within the EGT is that inclusive decision-making can resolve this tension (Van Assche et al., 2011; 2014). Other scholars believe that poor communication "encapsulates, in many ways, the challenges that communities living on the margins of mines face and the myriad ways in which they air grievances and engage authorities" (Mujere, 2015: 240).

The Bafokeng Land Buyers' Association (BLBA), which has a long-standing dispute with RBN over land rights and mineral royalties, articulates its concern as follows:

> The Royal Bafokeng Nation is known as one of Africa's success stories, for building a nation that benefits from the wealth beneath the land it owns. At the centre of this success is a king who cares for his people, and ensures that the "nation" is developed from the royalties coming from the mining businesses on their land.
>
> (BLBA, 2016)

BLBA further asserted that the media headlines do not often mention, if at all, is that the Bafokeng land is in dispute and that, in time, the success story could turn into a nightmare for all those who have bet on it should the courts decide that the story of the Royal Bafokeng. BLBA regards the RBN authority as a perpetuation of colonial-style conquest against defenceless communities (BLBA, 2016).

Hämäläinen and Lahtinen (2016) suggest a solution to this impasse, which, in part, is based on following different paths to resolve problem situations, with multiple decision-makers and stakeholders holding different preferences and views. This is particularly the case for RBN, which, besides being arguably the most transparent tribal authority in South Africa, is required to demonstrate repeatedly that it is indeed accountable to its subjects. However, as to be expected in an increasingly litigious society, as is the case in South Africa, the trust for RNB has eroded in recent years, particularly, over the integrity of its decision-making process. For example, in September 2005, the Supreme Council of RBN decided to institute legal proceedings against the Land Affairs Minister of South Africa. RBN was contesting that there was no trust relationship between it and the minister and it was the outright owner of over 60 properties, which, according to the title deeds, were being held by the minister "in trust for" the RBN (Ampofo-Anti, 2018; Pavlovic, 2018).

BLBA challenged the decision of the Supreme Council on the basis that it does not have the right to institute significant legal proceedings without consulting the community. Accordingly, in an outcome that Pavlovic (2018) refers to as "a constitutional endorsement for direct democracy in customary law," the court ruled in favour of BLBA:

> that it was part of the RBN's customary law that all matters of a 'public concern' had to be referred to broad consultation for the community to debate. The court, however, disagreed with the argument that the Kgosi had the sole right to determine which matters were of a "public concern" and needed to be referred to broader consultation. The court emphasised that customary law must be interpreted in light of the South African Constitution and its values, finding that public consultation and participation in decision-making are key components in promoting and strengthening democracy, and protecting rights and freedoms. Without a duty to consult the community, the community wouldn't have any ability to participate in the management of their assets. The court accordingly held that there is a duty under RBN's customary law to consult with the community on matters of public importance. The Supreme Council's failure to consult with the community regarding the legal proceedings meant that there was no valid decision to proceed with the court case.
>
> (Ampofo-Anti, 2018; Pavlovic, 2018)

Interestingly, Ampofo-Anti (2018) argues that the court finding dispels the claim that traditional leadership is incompatible with democratic principles of governance but highlights the unique manner, and path dependence, one can add, in which indigenous systems of governance accommodate principles of transparency, accountability, and consultation. Notwithstanding this, McKelvey et al. (2019) caution that governance is about understanding the different ways of developing rules and norms to interact and make decisions collectively. Challenges to RBN's historical growth-focused path dependency, which is "locked-in" to the tribal consultation system, raise a question about the future of such consultation pathways and attempts to ensure that residents have a greater sense of ownership. The challenge of the RBN's unilateral decision-making is not a case of community members seeking legal remedy against their tribal authority.

Another community in the North West Province of South Africa – under the Bagatla-Ba-Mosetla tribal authority in Makapanstad – restrained and interdicted their chief and headman from allocating and selling the tribal land (Jordaan, 2017). This explains why Leonard (2019) believes that traditional leadership structures pose significant barriers to the inclusion of local community concerns and decisions over mining development, or local community participation in decision-making processes, as well as "how political pluralism [as exuded by customary leadership] tribal is more than mere binary oppositions, rather is a series of complex and locally situated realities" (Cook, 2005: 125).

Broadly, this elucidates how institutional path dependence happens as a result of tribal actors who wish to block change in tribal consultative systems from taking place. However, it also demonstrates – as will be seen below – that direct community control of mineral wealth, as in the case of RNB, remains a major factor in communal resistance and socio-political conflict in tribal communities.

Local impacts and benefits in Phokeng

This section focuses on another key determinant of the social acceptability of mining projects: local impact and benefits. According to Owen &Kemp (2013), the promise of benefit to the community in the form of financial gains, as well as non-financial benefits, gives economic legitimacy to mining companies. Conventionally, the economic benefits of mining are measured in terms of variables such as employment and income (Measham & Zhang, 2019). In this study, the effects of mining activities on the community – in this case, RBN as the management authority of the revenue from mining – were considered on the two pillars of social licensing: the social and economic benefits of RBNDT.

At a time when several questions were raised about the resurgence of the traditional leadership form of governance, there are concerns about democratic inclusivity (Ubink, 2008). In the case of RNB – apart from the glossy and digital annual reports – a pertinent question is the extent to which the communal resources of the nation translate into individual well-being. In this regard, the quarterly performance report of RBN for the first quarter of 2018 indicates several achievements, which include projects rendered in kind by the Rustenburg Local Municipality, mining social and labour plans, and other strategic partnerships with an approximate value of R23 million. The report further revealed that a total of 11,280 youth (aged 18–35) benefited directly from interventions fostered by several units of RBN. The coordinated activities of RNBDT, which are focused on improving outcomes for youth, are further revealed by the formation of the first Bafokeng Student Lekgotla group and the first Bafokeng Student Conference, which was planned for the Youth Day Weekend in June 2018. The aim of these initiatives is to establish a mutually beneficial network of students who have been supported financially by RBNDT. In respect of employment, the RNB report states that 306 employment opportunities were created for the Bafokeng at the end of the first quarter through the combined efforts of local small, medium, and micro enterprises; Phokeng Mall (owned by RBN); and other units of RBN (RBN, 2018).

The CEO of the RBNDT further highlighted the importance of investing in the education of the Bafokeng. He indicated that RBN formalised their intervention in education since 2017 by establishing the Royal Bafokeng Institute. The institute offers technical, vocational, and educational training programmes that are aimed at post-school training for those youth who do not qualify for university enrolment. As he stated, "Our skills development training programmes are invaluable and, at times, more pertinent for many residents." He further mentioned that RBN piloted five early child development sites and partnered with the Department of Education to restore discipline at schools. In reiterating the Bafokeng's philosophy of managing risks prudently while having to deal with the rising, and at times, unrealistic expectations of the residents, the CEO stated:

> Constructive criticism of our operations are welcome, but there are too many arm-chair critics who fail to understand the daunting task of balancing act. Certainly, if we go the route of paying a monthly stipend to unemployed Bafokeng, we will be applauded for that. But how sustainable is such an approach? Unlike many other community trusts, we have deliberately avoided opulent lifestyle based on extensive consumerism. This bodes ill for any society. We understand the wide range of challenges the Bafokeng face, but we also take seriously the responsibility they entrusted us with to look after their wealth. While we leverage many sophisticated investment processes, there is a strong conviction that the wellbeing of the Bafokeng is the driving force behind our plans.

A group of young people and students who call themselves the Royal Bafokeng Diamonds and being in pursuit of Vision 2020 and MasterPlan 2035 shared the CEO's sentiments. However, the group took a rather critical view of colonialism and apartheid in South Africa:

> We are aware of the challenges that continue to haunt our 29 villages and five new established regions. The challenges are not new as they are experienced nationally across provinces. We are aware of the challenges of poverty, unemployment, illiteracy, underdevelopment and the youth. The challenge of the youth is of paramount importance in such a nation with international reputation and local acclaim. The perverse and idle tendencies that were reinforced by the malfunctioning societies and families during

colonialism and the rapid advance of the world towards development have left scars and marks that affect young people today. The lack of role models and sufficient and necessary education make young people susceptible to the corrupt forces of nature and human life. We have to admit that they will never find the light on their own without concern and constant engagement of societal leaders in their affairs.

(Royal Bafokeng Diamonds, 2016)

The CEO of RBNDT further explained that RNB operates, but we operate in an environment where people are not paying for basic services because they believe RBN has the financial muscle to sustain such malpractices. He believes that, in itself, this has been a huge benefit for residents in many villages for some time now. Attempts from RNB to let residents pay for basic services such as water and refuse collection have led to contestation in some areas. However, the CEO states that their budgetary constraints call for austerity measures in operational costs.

The magnitude of this concern of the CEO of RBNDT became very clear while interviewing residents. One male participant in the general focus group described how fortunate they have been "to live in one of the cleanest towns of blacks in South Africa." He further described that it is "such a good feeling to go the mall nearby and know that it belongs *go rona Batswana* [to us Tswanas]." A business owner said he moved to Phokeng 22 years ago and had no idea that he will stay for that long in the area. What impressed him was the honesty and level of professionalism of RBN when dealing with stakeholders such as business owners. "I have the utmost respect for the Bafokeng. They have responded to several requests from us in respect of local economic development and dealing with crime in the area," he continued. As a result, Phokeng has become a thriving model for sustainable community-based tourism and family-based agricultural projects. Further gains mentioned are typical of those found in abundant-resource communities, such as new markets for goods and services and the demand for local procurement.

Many other positive views about the RBN and the communal management of the nation's resources include the high number of jobs created for locals at the Royal Bafokeng Platinum and Bakubung mines (owned by RBN). Also, it is apparent that many are also satisfied with the positive welfare effects of mining in Phokeng. For example, many participants emphasised the improved accessibility of health-care services and access to information. They were adamant that Phokeng used to be just another black residential area in South Africa. Through the relentless efforts of the Bafokeng tribal leadership, Phokeng has received world acclaim for how it manages community resources. However, the greatest praise for the RNB model came a few years ago from Modipa (2013), when he appraised land restitution which has restored land ownership and the potential economic benefits to these new landowners in South Africa. About RNB, Modipa (2013) explains:

> Today, Bafokeng operates within the parameters of their Vision 2020 and the Masterplan. The Bafokeng commercial activities yield a surplus of

between R400 million and R1 billion per year. This surplus is not used to create an opulent lifestyle for the King. Most senior managers and executives in Bafokeng entities earn far more than the King. Will the King possibly demand a higher salary and opulent fringe benefits? No, no, he will never do that. His is a service to his people. You cannot associate the phrase 'financial profligacy' with him. He does not live in some exotic palace; he lives in a normal house with normal furniture. His house is in Legato (royal estate) and it is not even called a Palace. The makeshift offices in Legato where the King and the Queen Mother work are not at all attractive. The commercial returns are used to build infrastructure and to provide social services to the community. It is important to state that the bulk of the surplus is used in the community's educational endeavours. If we are serious about economic empowerment in SA, these are some of the things other traditional communities can learn from Bafokeng. Bafokeng was able to get financial support from banks because they have well-defined community and commercial governance structures.

I also have to add that during several of my visits to Phokeng over the three years of fieldwork, I was struck by the cleanliness of the area. In the field observations, I noticed that many houses had beautiful gardens, roads were tarred, streetlights were working in the evening, and there were no shacks (informal corrugated structures that are often built in informal settlements or behind formal structures in South Africa).

There were also several other participants of the study who believe that RNB has not contributed significantly to the standard of living in Phokeng and that they are personally not better off financially as a result of RNB or its mines. These participants tended to have less trust in the ability of RBN to act responsibly or do the right thing for the community of Phokeng. A common observation of these participants was, as in the case of Marais et al. (2018), a general lack of access to markets, increased costs of municipal services, and an exclusion from employment and other opportunities based on ethnicity.

Examples of these negative sentiments provided by the participants included "You have to be from the Bafokeng clan to land a job here"; "shareholders, and not the community reap the benefits from RNB"; and "adequate incomes, affordable housing, and food security is under threat as a result of RNB and the mines around us." Another one interjected:

> but what achievements are you referring to? Fancy buildings and malls mean nothing to some of us. We want to see tangible benefits. For a nation that is worth billions, the level of hopelessness and poverty in Phokeng is unacceptable.

Thus, for many of these participants, the success of RBN has failed to meet expectations of a better life. Earlier reported sentiments also confirm the concerns raised by the study. For example, Carte (2007) reported on a resident who stated:

> I am a member of the Bafokeng Nation and their things that never stop to amaze me about the authorities or His Majesty, in particular, it is true that majority of Bafokeng people are still living in severe poverty with no access to water and sanitation. The wealth of Bafokeng seems to benefit the few minority and the Royal House. But what worries me most is Kgosi's lack of confidence in the local companies, most of the business opportunities are given to Johannesburg companies not that the local companies do not have capacity but because they think local companies cannot deliver. But also it goes back to the same authority for their inability to develop capacity of those companies because there is a special unit that focuses on local economic development. Until Kgosi Leruo Molotlegi believes on his own people, the wealth of Bafokeng will never mean anything to the ordinary young mofokeng.

Another added:

> It is evident that, while platinum mining has benefited mining capital and elites such as the Royal Bafokeng Nation [RBN], the wealth has failed to trickle down to the lower classes. Yet [it] is the lower classes especially the workers, villagers and residents of informal settlements that have emerged on the margins of the mines who bear the brunt of the worst effects of mining.
> (Evans, 2015)

According to the survey, most residents of Phokeng have negative perceptions about the procedural fairness and general trust of RBDNT (see Table 4.2). In respect of procedural fairness, 59% of the respondents do not believe that RBNDT listens to and respect their opinions. A slighter higher proportion of respondents (63%) indicated that they do not believe that RBNDT is prepared to change its practices in response to community sentiment.

Table 4.2 Procedural fairness and trust in RBNDT (percentage in parentheses)

Variable	Yes	No	Total[a]
Procedural fairness	77 (41.5)	108 (58.5)	185
I believe RBNDT listens to and respects our opinions	68 (36.7)	117 (63.3)	185
I believe RBNDT is prepared to change its practices in response to community sentiment			
Trust in RBNDT			
I am able to trust RBNDT to act responsibly	137 (74.2)	58 (25.8)	185
I trust RNBDT to act in the best interest of the community	74 (39.9)	111 (60.1)	185
My trust in RNBDT has decreased over the past five years	92 (50.8)	89 (49.2)	181[b]

a A total of 185 who either knew little or extensively about RNBDT.
b Four who did not know were excluded.

100 Royal Bafokeng Nation

Table 4.3 RNBDT's contribution to the quality of life of Phokeng residents (percentage in parentheses)

Variable	Yes	No	Total
Gender			
Male	38 (35.2)	70 (64.8)	108
Female	28 (36.4)	49 (63.7)	77
Age			
21–35	12 (20.7)	48 (79.3)	60
36–55	37 (43.5)	49 (56.5)	86
56 and older	25 (66.4)	14 (33.6)	39
Educational qualification			
No formal schooling	6 (66.7)	3 (33.3)	9
Primary school (Grades 1–7)	20 (38.1)	32 (61.9)	52
High school (Grades 8–12)	28 (28.6)	70 (71.4)	98
Postgraduate – diploma	6 (45.3)	8 (54.7)	14
Postgraduate – degree	8 (71.0)	4 (29.0)	12

Respondents were also asked whether they believe that RNBDT makes a major contribution to the quality of life of Phokeng residents. Overall, 64% of the respondents do not believe that RNBDT makes a major contribution to the quality of life of Phokeng residents. There was no statistical significance between genders, as 64.8% and 63.7% of males and females, respectively, do not believe that RNBDT makes a major contribution to the quality of life of Phokeng residents (Table 4.3).

However, there were varied perceptions in respect of age and level of education about the contribution of RBNDT. While 79% of the young respondents and 56% of the middle-aged have negative thoughts about RNBDT's contribution to the overall quality of life of residents, only a third of the older respondents shared similar thoughts. Interestingly, in respect of education, respondents with no formal education (67%) and those with a tertiary degree (71%) are most likely to believe that RBNDT contributes to the overall quality of life of Phokeng residents.

Future pathways to social licensing

Scholars have discerned the gradual evolution of citizens' rights, more broadly, and mining-affected communities' rights to decide about developments in their locality, in particular. However, almost two decades ago (and still pertinent today), many liberal democratic theories expressed reservations about the viability of systems based on widespread political participation or the universal acceptance of participation as a key element in a successful development strategy (Macdonald, 2001). Given the unanimity about the success of RBN and its portrayal as an inherently participatory entity due to its expanded tribal consultative systems (as reported in its various print and digital reports, and confirmed by the CEO of RBNDT) and the great variance in the views of residents of Phokeng about RNB, this section broadly focuses on the future: possible solutions and behavioural intentions.

According to the CEO of RBNDT, the future looks bright. However, several challenges are pertinent. RNB, in general, is challenged to seek ways to do more with less, but its partnership with education institutions and its vision for the education of the Bafokeng nation remain its flagship projects. He emphasises that the focus on education is part of Vision 2035, the nation's long-term strategic plan to address challenges by focusing more on physical and infrastructure development. According to Moumo Integrated Developments (2016), the objective of the plan is to create sustainable growth and prosperity for the community and to realise the potential value of the land, which is the single largest asset for the Bafokeng community. In particular, the plan takes into consideration socio-economic factors, and respect for the culture, history, and heritage of the community to achieve four specific objectives. First, it aims to ensure a competitive and vibrant economy that is supported by a well-educated and skilled workforce. Second, it aims to ensure a high quality of life for the RBN residents that is comparable with that of a suburban township in a developed country. Third, it aims to embrace efficient transportation network and infrastructure that support the activities of residents and businesses. Finally, it aims to ensure that the traditional culture and heritage of the Bafokeng tribe continues to bind the social fabric of the community for future generations (Moumo Integrated Developments, 2016).

The CEO of RBNDT noted that the recent discourse on land expropriation in South Africa and the over-dependence on mining revenue point to the need to find new ways of responding to and solving challenges. This, he continued, needs the intersection of RBN, the broader public sector, and the mining industry. However, this is often compounded by institutional inertia that prevents new ways of working. For RBNDT, there is a need to adopt a set of unique approaches and to forge new partnerships. The CEO mentioned as an example how students who have been financially assisted by RBNDT would rather go and look for a job in Gauteng Province instead of starting a business in Phokeng. This is a huge barrier for RBN, as several companies, such as Volkswagen, wanted to open their plants in Rustenburg, but they often ask RBN about the skills base in the region. "These are opportunities that we need to harness at all costs," the CEO said.

The literature emphasises that that complexity impacts path dependence. In this regard, Koch et al. (2009: 79) states that "path-dependence results from poor decisions that are because people in highly complex situations tend to neglect future developments at the expense of information on present situations." This concern is reflected by the CEO of RBNDT, who said new costs arose when RNB assumed responsibility for basic services, including health and social services, in the areas under its jurisdiction. A major concern for the RBNDT is that the changing economic realities in South Africa, as elsewhere, have a huge impact on RBN. If not properly managed, these conditions may affect RBN's ability to provide services on behalf of the municipality. However, another wider concern about the provision of basic services is the potential lack of clarity on whether the RBN or the Rustenburg Local Municipality should provide services. For example, an earlier study titled "Platinum, Poverty and Protests" highlights how the residents of Luka village (under the administration of RBN) have resorted to insurgent practices to force the RBN, the Rustenburg Local Municipality, and platinum mining companies to provide them with basic amenities and sources of livelihood (Mujere, 2015).

The CEO was satisfied that Phokeng had not experienced any protests for more than two years before the interviews. However, he emphasised that the political landscape in South Africa has changed dramatically and so has citizen activism to assert the rights of the disadvantaged. This is seen by the increasingly high number of land tenure challenges. In conclusion, the CEO of RBNDT recognised the need for the "continued prudent management assets and resources, while focusing on the well-being of distressed communities."

As the preceding responses reveal, it is a challenge to deal sustainability issues in natural-resource management, and finding ways to exploit mining benefits and minimise impacts from boom-bust dynamics. When asked about the future of RNB and behavioural tendencies, the majority of the participants recognised the legitimacy and inherent importance of RNB, but still also felt that improvements are needed in how RNB engages the community and distributes economic benefits. The interviewed residents held positive attitudes towards the potential of RNB to fulfil the expectation of meeting the basic needs of residents. A future beneficial development pathway for some is when "real benefits in the form of improved livelihoods become realistic for residents." For others, "the presumption of a successful RBN must lead to more democratic decision-making power for residents."

The survey of RNB residents gauged the future behavioural intentions of the interviewed residents of Phokeng (Table 4.4) and showed that almost two-thirds of the respondents (61.4%) expressed that they were likely to complain to people about RBNDT. A fairly high proportion of respondents reported that it is unlikely that they will attempt to discourage others from working with RBNDT (80.3%) and encourage others to protests against RBNDT (62.9%). Furthermore, 43.5% indicated that it is likely that they will participate in protests by residents against RBNDT. The vast majority of the respondents (76.8%) expressed that it is unlikely that Phokeng will be a better place to live in a decade from the time of the survey.

Table 4.4 Future behavioural intention regarding selected variables (percentage in parentheses)

Variable[a]	Likely	Unlikely	Total
How likely will you complain to people about RBNDT?	114 (61.4)	71 (38.6)	185
How likely will you attempt to discourage others from working with RNDT?	36 (19.7)	149 (80.3)	185
How likely will you encourage others to protests against RBNDT?	69 (37.1)	116 (62.9)	185
How likely will you participate in a protest by residents against RBNDT?	81 (43.5)	104 (56.5)	185
How likely will your community be a better place to live in a decade from now?	43 (23.2)	142 (76.8)	185

[a]Measured with a seven-point Likert scale.

While North West Province has experienced several labour-related protests, including the killing of 34 striking miners who were demonstrating against the London-based Lonmin company in Marikana in August 2012 (Alexander, 2013; Botiveau, 2014; Hayem, 2016), RNB has experienced no protests in Phokeng. Path dependency helps us understand that structural opportunities for social mobilisation are more stable features of political institutions and culture – like the RBN governance system – that change only gradually over decades. We are reminded by Gramson and Meyer (1996: 222) that structural opportunities for political participation "are essentially fixed and given, barring dramatic and unforeseen changes beyond their control." Still, beyond the unprecedented economic growth of RNB, the dissatisfaction of residents about its consultation processes and recurring legal disputes over land rights has the potential to disrupt RNB's pathways towards an SLO and sustainable future for its subjects.

Discussion – the significance of path dependency

This study analysed the case of RNB, which is renowned as a model for the prudent community-based management of revenues from natural resources. The findings of the study resonate strongly with the observations of previous studies on RNB. However, it offers insight into two contested issues: the primary factor behind the success of RNB and its consultation processes.

The study demonstrates how, over more than a century and eight decades, the RNB sphere of management has come to encompass dynamic policy pathways, reconciling traditional and modern governance institutions (Mbenga & Manson, 2010; Cook, 2011; Flomenhoft, 2018). This approach has worked for RNB, which, like any other organisation, had to deal with structural mechanisms, market incentives, and the planning measures of the colonialists and the apartheid government in South Africa. Throughout this journey, RNB remains an ardent adherent to its customary practices and, through its visionary leadership (Mnwana, 2014), managed to navigate through the vulnerabilities and opportunities of mining (Haasnoot et al., 2013).

While the proponents of the EGT suggest that institutions are "locked-in" to past decisions (Beunen et al., 2016), the RNB case demonstrates that the impeccable leadership acumen of the Bafokeng leaders is behind the success of the tribal authority. The unique platinum abundance in Rustenburg strengthened RNB's asset base. However, I argue that attributing the success of RNB solely to its resource abundance obscures the role of institutions. Bridges (2017: 169) laments the point that institutions are critical in efforts to hasten the "adoption of sustainability ideals and in the implementation of associated projects." There will always be a chasm between customary and modern practices in many respects. However, for the Bafokeng leadership to have been able to exercise due diligence over many generations is commendable. It is strategic planning alongside the calculated decision to reconcile tribal and modern institutions of governance that are, in part, behind the success of RNB. Thus, history did matter in the case of RNB. Moreover, that is a history in which wise stewardship over non-replenishable natural resources was exercised instead of the route of consumerism taken by many other communities who receive revenues from mining. However, the question remains whether they will be able to maintain their social acceptance by the Bafokeng nation.

This brings me to the second factor raised by the study: community consultation. It is widely accepted that the processes mining companies use to engage and build trust with local communities are central to the granting and maintaining of a social licence to operate, on the one hand. On the other hand, the lack of benefits on the social and economic infrastructure, as well as the community members' perception of the quality of the consultation, for example, significantly affects the "community's acceptance of the mining operation through inferred trustworthiness of the company" (Moffat & Zhang, 2014: 61). Similarly, close collaboration, respect, and trust are necessary building blocks in facilitating responsible mineral development that enhances benefits for local communities (Canada's Minerals and Metals Sector, 2016). However, the study suggests that RNB's public consultation processes, including *Kgothakgothe*, provide limited space for the broader citizens of Phokeng to exercise their democratic rights over the management of the communal resources.

The study further demonstrates that the presence of consultation mechanisms should not be equated with effective participation or benefits for the community. The aim of community-based natural-resource management is to maximise public benefit derived from mining royalties. Several studies have found that traditional leaders continue to exercise complete and sole authority over revenues from natural resources and use this capital to entrench their positions (Bennet et al., 2013; De Kadt & Larreguy, 2018; Leonard, 2019). It is thus no surprise, as the study shows, that there is little evidence of a broader participation of residents of Phokeng in control over the decision-making processes of RNB, thus calling for the opening of spaces of engagement.

Conclusion – inclusive pathways towards a sustainable future

Applying path dependence to community-based natural-resource management provides insights into how lock-in mechanisms influence different path dependencies and thus social licensing. The case of RNB is instructive for demonstrating how successive generations of leaders and their actions underpinned by the Tswana customary values created a deeply path-dependent system of managing the benefits accrued from its land tenure. This case allows us to understand that the current success of the tribal authority is the product of its history, including its intelligentsia and visionary leadership. The study confirms some of the conclusions about how history influences path dependencies. For example, it is evident that, throughout its protracted history, RNB – embedded into tribal pathways of decision-making – found the most suitable solution to problems in pursuit of its strategies.

The RNB case further reinforces the point made by leading scholars concerning the implementation of decision-making pathways which need to find expression at the community (individual or household) level (Van Assche et al., 2014; Beunen et al., 2016; Parsons et al., 2019) and the importance of building trust with communities to maintain an SLO (Mofatt & Zhang, 2014). However, societies like RBN that have been influenced by path-dependent processes are also trapped by historical contingencies.

Thus far, RNB has achieved considerable success by reconciling traditional institutions of governance with modern governance institutions. This process locked-in RNB in the practice of systematically engaging long-term planning and thereby protected the very foundation of its tribal values. Such planning, including Vision 2035 of RNB, "highlights clear contingency, and is characteristic of path dependency" (Parsons et al., 2019). However, following the mounting conflict over the management of natural resources, in general, and community-based natural-resource management by traditional authorities, in particular, RNB will have to disrupt its traditional consultative system by opening it up to the broader democratic civic engagement.

Another area that RNB should be keen to focus on is to meet the demands and expectations of the Bafokeng nation. Public goods, such as education, health, water, sanitation, and housing, are indispensable assets. Therefore, neo-liberal strategies and institutions are good in economic terms but do not translate into public good for residents of Phokeng. Consequently, the loss of confidence of Phokeng residents in RNB's ability to improve their economic conditions may risk the harmony in this area. However, mining does not only have negative consequences for the environment but also for existing cultural traditions. RNB has received scant credit for its resilience in pursuing structures and processes ostensibly designed to address local adverse and beneficial social and economic impacts arising from mining while preserving the Bafokeng cultural tradition and norms (this is what any great nation across the world aspires to do). This assertion is confirmed by Wilson (2016: 73) who notes, "the success of efforts to establish a social licence that benefits all parties depends on local expectations and historical experience in particular sociocultural and political contexts."

Nevertheless, I believe that this is where RNB has to ensure it maintains its social acceptability by proceeding with its vision, but most importantly, to balance its vision with its goal of fulfilling its social responsibility duties towards residents. This can be achieved by reconciling the traditional consultative institutions with modern principles of participatory governance. Whether and to what degree the RBNDT will seek to create inclusive pathways towards a sustainable future for all the Bafokeng remains to be seen.

References

Alexander, P. (2013). Marikana, turning point in South African history. *Review of the African Political Economy* (December): 605–619.
Ampofo-Anti, A.O. (2018). Court orders tribal authority to act democratically. *GroundUp* [online]. Available at www.groundup.org.za/article/court-orders-tribal-authority-act-more-democratically/ (accessed 29 December 2018).
Bench Marks Foundation (2007). The policy gap. Review of the corporate social responsibility programs of the platinum mining industry in the platinum producing region of

the North West Province [online]. Available at www.bench-marks.org.za/research/Rustenburg%20platinum%20research%20summary.pdf (accessed 4 January 2018).

Bennet, J., Ainslie, A. & Davis, J. (2013). Contested institutions? Traditional leaders and land access and control in communal areas of Eastern Cape Province, South Africa. *Land Use Policy*, 32(May): 27–38.

Beunen, R., Van Assche, K. & Duineveld, M. (2016). *Evolutionary governance theory*. New York: Springer.

Cawood, F.T. & Minnitt, R.C.A. (1998). A historical perspective on the economics of the ownership of mineral rights ownership. *Journal of South African Institute of Mining and Metallurgy*, 98: 369–370.

BLBA (Bafokeng Land Buyers' Association) (2016). A section of a tribe [online]. Available at http://bafokeng-communities.blogspot.com/ (accessed 12 May 2018).

Botiveau, R. (2014). Briefing: The politics of Marikana and South Africa's changing labour relations. *African Affairs*, 113(450): 128–137.

Bridges, A. (2017). The role of institutions in sustainable urban governance. *Natural Resources Forum*, 40 (2016): 169–179.

Canada's Minerals and Metals Sector (2016). Good practices in community engagement and readiness (2nd edn) [online]. Available at www.nrcan.gc.ca/sites/www.nrcan.gc.ca/files/mineralsmetals/files/pdf/rmd-rrm/GoodPractices2ed_En.pdf (accessed 04 April 2019).

Carte, D. (2007). The barefoot Bafokeng billionaires [online]. Available at www.moneyweb.co.za/archive/the-barefoot-bafokeng-billionaires/ (accessed 12 January 2019).

Coertze, R.D. (1990). *Bafokeng family law and law of succession*. Pretoria: Sabra.

Comaroff, J.L. & Comaroff, J. (2009). *Ethnicity Inc*. Chicago: University of Chicago Press.

Cook, S.E. (2005). Chiefs, kings, corporatization and democracy: A South African case study. *Brown Journal of World Affairs*, 12(1): 125–137.

Cook, S.E. (2009). *The business of being Bafokeng: The corporatisation of a tribal authority in South Africa*. Paper presented at the Wenner-Gren International Symposium No. 140 on Corporate Lives: New Perspectives on the Social Life of the Corporate Form, held in conjunction with the School for Advanced Research, Santa Fe, NM, 21–27 August.

Cook, S.E. (2011). The business of being Bafokeng: The corporatization of a tribal authority in South Africa. *Current Anthropology*, 52(S3): S151–S159.

Cook, S.E. (2013). Community management of mineral resources: The case of the Royal Bafokeng Nation *Journal of the Southern African Institute of Mining and Metallurgy*, 113(1): 61–66.

Corruption Watch. (2019). Mining royalties research report 2018 [online]. Available at www.corruptionwatch.org.za/wp-content/uploads/2019/03/Mining-royalties-research-report-final1.pdf (accessed 30 June 2019).

De Kadt, D. & Larreguy, H.A. (2018). Agents of the regime? Traditional leaders and electoral politics in South Africa. *The Journal of Politics*, 80(2): 382–399.

Evans, S. (2015). Behind community protests in the platinum belt. *Mail & Guardian* [online]. Available at https://mg.co.za/article/2015-08-05-who-will-bear-the-brunt-of-mining-communities (accessed on 24 November 2017).

Flomenhoft, G. (2018). Historical and empirical basis for communal title in minerals at the national level: Does ownership matter for human development? *Sustainability*, 10(6): 1958.

Flomenhoft, G. (2019). Communal land and the attitudes of the Bafokeng on benefits from mineral rights. *Journal South African Journal of International Affairs*, 26(2): 277–301.

Gaborone, G.T. (2014). *Lebone II College of the Royal Bafokeng: Building a Mofokeng who can actively participate in South Africa's 21st century economy*. Dissertation. Johannesburg: Wits University [online]. Available at http://wiredspace.wits.ac.za/bitstream/handle/10539/15556/Research%20Report%200617122k.pdf?sequence=2&isAllowed=y (accessed 3 January 2018).

Gramson, W.A. & Meyer, D. (1996). Framing political opportunity. In: D. McAdam, J.D. McCharthy & M.A. Zald (eds). *Comparative perspectives on social movements: Political opportunities, mobilizing structures, and cultural framings* (pp. 275–311). Cambridge: Cambridge University Press.

Haasnoot, J., Kwakke, J.H., Walker, W.E. & Maat, J. (2013). Dynamic adaptive policy pathways: A method for crafting robust decisions for a deeply uncertain world. *Global Environmental Change*, 23(2): 485–498.

Hämäläinen, R.J. & Lahtinen, T.J. (2016). Path dependence in operational research – How the modeling process can influence the results. 3: 14–20.

Hull, S. (2018) [online]. Available at https://theconversation.com/why-giving-south-africans-title-deeds-isnt-the-panacea-for-land-reform-98106 (accessed 14 October 2018).

Jordaan, N. (2017). North West community want their chief arrested for 'unlawfully' selling their land [online]. Available at www.timeslive.co.za/news/south-africa/2017-08-29-north-west-community-want-their-chief-arrested-for-unlawfully-selling-their-land/ (accessed 19 June 2018).

Khunou, F.S. (2011). Traditional leadership and governance: Legislative environment and policy development in a democratic South Africa. *International Journal of Humanities and Social Science*, 1(9): 278–290.

Koch, J., Eisend, M. & Petermann, A. (2009). Path dependence in decision-making processes: Exploring the impact of complexity under increasing returns. *BuR Business Research Journal*, 2(1): 67–84.

Lacey, J., Carr-Cornish, S., Zhang, A., Eglinton, K. & Moffat, K. (2017). The art and science of community relations: Procedural fairness at Newmont's Waihi Gold operations, New Zealand. *Resources Policy*, 52: 245–254.

Leonard, L. (2019). Traditional leadership, community participation and mining development in South Africa: The case of Fuleni, Saint Lucia, KwaZulu-Natal. *Land Use Policy*, 86(July): 290–298.

Lucas, A. (2016). Elite capture and corruption in two villages in Bengkulu Province, Sumatra. *Human Ecology*, 44(3): 287–300.

Macdonald, L. (2001). NGOs and the discourses of participatory development in Costa Rica. In: H. Veltmeyer & A. O'Malley (eds). *Transcending neoliberalism: Community-based development in Latin America*. Bloomfield, NJ: Kumarian Press, pp. 125–153.

Manson, A. & Mbenga, B. (2003). 'The richest tribe in Africa': Platinum mining and the Bafokeng in South Africa's North West Province, 1965–1999. *Journal of Southern African Studies*, 29(1): 25–47.

Matebesi, S.Z. & Marais L. (2018). Social licensing and mining in South Africa: Reflections from community protests at a mining site. *Resources Policy*, 59(December): 371–378.

Marais, L., Van Rooyen, D., Lenka, M., Cloete, J., Denoon-Stevens, S., Mocwagae, K., Jacobs, M. & Riet, J. (2018). The background to the Postmansburg study. In: P. Burger, L. Marais & D. Van Rooyen (eds). *Mining and community in South Africa: From small town to iron town*. New York: Routledge.

Maseko, M.J. (2015). *The transformation of traditional leadership: A case study of the Simdlangentsha Traditional Council and its relationship with local government*. Doctoral thesis, Potchefstroom

Campus of the North-West University [online]. Available at https://pdfs.semanticscholar.org/c011/a837b7da8a6fcb581cd0e4a7e631eb711357.pdf (accessed 12 July 2018).

Mbenga, B. & Manson, A. (2010). *'People of the dew': A history of the Bafokeng of Phokeng-Rustenburg Region, South Africa, from early times to 2000.* Johannesburg: Jacana.

McKelvey, M., Zaring, O. & Szücs, S. (2019). Conceptualizing evolutionary governance routines: Governance at the interface of science and technology with knowledge-intensive innovative entrepreneurship. *Journal of Evolutionary Economics.*

Measham, T.G. & Zhang, A. (2019). Social licence, gender and mining: Moral conviction and perceived economic importance. *Resources Policy*, 6(June): 361–368.

Mnwana, S.C. (2012). *Participation and paradoxes: Community control of mineral wealth in South Africa's Royal Bafokeng and Bakgatla Ba Kgafela communities.* Fort Hare: University of Fort Hare.

Mnwana, S.C. (2014). Mineral wealth – 'in the name of morafe'? Community control in South Africa's 'Platinum Valley'. *Development Southern Africa*, 3(6): 826–842.

Mnwana, S.C. & Akpan, W. (2009). *Platinum wealth, community participation and social inequality in South Africa's Royal Bafokeng community – A paradox of plenty?* 4th International Conference on Sustainable Development Indicators in the Minerals Industry. Australasian Institute of Mining and Metallurgy, Proceedings SDIMI, pp. 283–290.

Modipa, M.E. (2007). *Sustainable socio-economic development and poverty alleviation strategies for communities: A review of the Royal Bafokeng Nation initiative.* Thesis. Durban: University of KwaZulu Natal.

Modipa, M. (2013). Lessons from the Bafokeng [online]. Available at www.politicsweb.co.za/news-and-analysis/lessons-from-the-bafokeng (accessed 12 June 2017).

Moffat, K. & Zhang, A. (2014). The paths to social licence to operate: An integrative model explaining community acceptance of mining. *Resources Policy*, 39(March): 61–70.

Moumo Integrated Development (2016). Royal Bafokeng National Master Plan [online]. Available at www.moumo.co.za/about-us/master-plan.html (accessed 12 June 2017).

Mrema, J.P. (2017). Tanzania (Chapter 10). In: A. Williams, A. & P. Le Billion, P (eds). *Corruption, natural resources and development* (pp. 131–141). New York: Edward Elgar Publishing, Chapter 10, pp. 131–141.

Mujere, J. (2015). Labour, capital and society/travail. *Labour, Capital & Society*, 48(1&2): 240–266.

Nthau, N.G. (2002). *The possible role of traditional leadership in development planning in South Africa: A case study of the Bafokeng Community in North West Province.* Thesis. Johannesburg: University of the Witwatersrand.

O'Faircheallaigh, C. (1998). Resource development in indigenous societies. *World Development*, 26(3): 381–394.

Owen, J. & Kemp, D. (2013). Social licence and mining: A critical perspective. *Resources Policy*, 38(1): 29–35.

PMG (Parliamentary Monitoring Group) (2014). Roles and functions of traditional leaders: Department of Traditional Affairs Briefing [online]. Available at https://pmg.org.za/committee-meeting/17800/ (accessed 24 November 2018).

Parsons, M., Nalau, J., Fisher, K. & Brown, C. (2019). Disrupting path dependency: Making room for indigenous knowledge in river management. *Global Environmental Change*, 56: 95–113.

Pavlovic, C. (2018). A constitutional endorsement for direct democracy in customary law. *Mining Engineering News* [online] Available at https://m.engineeringnews.co.za/article/a-constitutional-endorsement-for-direct-democracy-in-customary-law-2018-06-18/rep_id:4719/company:hogan-lovells-2017-07-07 (accessed 4 January 2019).

Platteau, J. & Gaspart, F. (2003). The risk of resource misappropriation in community-driven development. *World Development*, 31(10): 1687–1703.

Royal Bafokeng Diamonds (2016). Royal Bafokeng Diamonds manifesto [online]. Available at http://royalbafokengdiamonds.cfsites.org/custom.php?pageid=39656 (accessed 14 June 2018).

RBN (Royal Bafokeng Nation) (2018). Royal Bafokeng Nation's quarterly performance report: 1st quarter of 2018 [online]. Available at www.rbnoperationsroom.com/files/display/id/14645 (accessed 23 November 2018).

Rustenburg Local Municipality (2019). Integrated development plan: Final IDP Review 20119/2020 [online]. Available at www.rustenburg.gov.za/Docs/Final_IDP_Review_2019-2020.pdf (accessed 24 November 2018).

Slack, K. (2009). Mining conflicts in Peru: Condition critical [online]. Available at www.oxfamamerica.org/newsandpublications/publications/briefing_papers/mining-Conflicts-in-peru-condition-critical/Mining-Conflicts-in-Peru-Condition-Critical.pdf (accessed 12 March 2018).

Tekwa, E.W., Fenichel, E.P., Levin, S.A. & Pinsky, M.L. (2019). Path-dependent institutions drive alternative stable states in conservation. *PNAS*, 116 (2): 689–694.

Thompson, L.F. (2015). *The Royal Bafokeng: A case study for the resource curse*. Dissertation. Cape Town: University of Cape Town.

Tieleman, J. & Uitermark, J. (2018). Chiefs in the city: Traditional authority in the modern state. *Sociology* (December 6).

Ubink, J. (2008). *Traditional authorities in Africa: Resurgence in an era of democratisation*. Amsterdam: Eiden University Press.

Van Assche, K., Beunen, R., Duineveld, M. & de Jong, H. (2013). Co-evolutions of planning and design: Risks and benefits of design perspectives in planning systems. *Planning Theory*, 12(2): 177–198.

Van Assche, K., Beunen, R., Jacobs, J. & Teampau, P. (2011). Crossing trails in the marshes: Rigidity and flexibility in the governance of the Danube Delta. *Journal of Environmental Planning and Management*, 54(8), 997–1018.

Van Assche, K., Beunen, R. & Duineveld, M. (2014). *Evolutionary governance theory: An introduction*. Heidelberg: Springer.

Walker, W.E., Rahman, S.A. & Cave, J. (2001). Adaptive policies, policy analysis, and policy-making. *European Journal of Operational Research*, 128(2): 282–289.

Williams, A. & Le Billion, P. (2017). *Corruption, natural resources and development*. New York: Edward Elgar Publishing.

Wilson, E. (2016). What is the social license to operate? Local perceptions of oil and gas projects in Russia's Komi Republic and Sakhalin Island. *Extractive Industry and Society*, 6(3): 73–81.

5 Social mobilisation against community development trusts in South Africa

Introduction

This chapter examines the process for achieving locally based social licences through community development trusts (CDTs) in locations where much conflict has been experienced in the Free State, Limpopo, and Northern Cape provinces of South Africa. Social licensing can play a significant role in bridging the gulf and enhancing relations between industry, government, and society (Prno & Slocombe, 2012; Bice, 2014; Koivurova et al., 2015; Owen & Kemp, 2014, 2017). Despite the advent of the Mineral and Petroleum Resources Development Act (MPDRA) of 2002 and subsequent court judgements that, among others, asserted the obligation of mining companies to consult with interested and affected parties, the widespread violence in mining-affected communities demonstrates the underlying concerns of civil society about being ignored in the participation in the community-based management of revenues from local mining projects.

Scholars suggest that while governments sought to establish strong domestic regulation of mining activities to protect and champion the interests of communities, these regulations are neither respected nor enforced (Centre for Applied Legal Studies, 2017; Mandela Institute, 2017; Marais et al., 2008; Matebesi & Marais, 2018). As in the case of Peru (and elsewhere in the world), the sporadic, isolated events of weak protest grassroots groups have been ineffectual at accelerating democratic responsiveness (Arce, 2014). This situation exemplifies the autonomous power enjoyed by, most often, international mining companies in the neoliberal world order (Marshall, 2015). Interestingly, the abuses by international mining companies have not gone unnoticed in their home countries, albeit from a community perspective.

Establishing new regulatory measures to hold mining companies to account for their activities abroad has long been a concern of civil society groups (Marshall, 2015). In May 2019, national human rights, faith, and labour organisations, along with concerned Torontonians, rallied in Toronto to call for an ombudsperson with the power to investigate corporate abuses abroad. In this regard, a lawyer who is representing Mayan Q'eqchi' people in Guatemala against Canadian mining giant Hudbay Minerals and its subsidiary HMI Nickel, Inc. (quoted in MiningWatch Canada, 2019) noted, "This is a classic bait-and-switch by the government – promise one thing, deliver another." He further explained, "It is pathetic that in 2019, not only do Canadian companies continue to export human rights and environmental abuses but worse, the Canadian government refuses to do anything that will address the problem" (MiningWatch Canada, 2019).

Prior research on evolutionary governance theory (EGT) highlights the importance of participation in democratic decision-making processes affecting the ability to build trust with communities (Van Assche et al., 2014). Recent studies on CDTs in South Africa have found several reasons for the widespread violent protests over how mineral royalties are managed through CDTs. These reasons include flawed mining royalty systems, involving greed, competition for a finite financial resource, deliberate exploitation, and mismanagement of funds and resources (Corruption Watch, 2018) and "a systemic lack of accountability and transparency by traditional leaders and mining companies towards ordinary rural citizens" (Pickering & Nyapisi, 2017: 27). Earlier international experiences also found that CDTs (or foundations) are presented publicly as an instrument to promote unstainable development and a more equitable relationship between companies and communities. In reality, they serve primarily to assimilate opposition (CERLAC, 2002: 18–19). Corruption Watch (2018) emphasises that these challenges of mining royalties system are "locked into" socio-historical and political paths. Furthermore, as in the case of Peru (and elsewhere in the world), the response of weak protest grassroots groups triggers sporadic and isolated events (Arce, 2014), which have become more brazen and violent.

In this chapter, I explore three CDTs which have drawn intense opposition and conflict from their communities. Five main observations can be made from the cases. First, they have experienced a high incidence of protest, though this varies considerably in respect of intensity. Second, the board of trustees is based on partisan connections. Third, they represent a toxic mix of maladministration and a penchant for offering incentives to select community activists. Fourth, that community activist groups, though perceived to be weak, have great influence in shaping social licences to operate (SLOs). Finally, and most importantly, the cases demonstrate that "effectively, mining areas are locked into pathways of conflict that persist irrespective of the institutional arrangements that have been devised to address this reality" (Matebesi & Marais, 2018).

The findings presented in this chapter show resonance in studies conducted on different contexts, for example, Kirsch's (2014) and Bebbington and Bury's (2013) studies on resource booms and busts in Latin America. Other scholars who made significant contributions to the disruptive capacity of the "subsoil" are Bunker (1989) and Boyd et al. (2001).

In the next section, I present the local contexts of the three cases. I then proceed to present and explain the three case studies that were undertaken to inform this book. The section presents these cases individually according to the main question I am interested in how particular events in the cases shape the different paths to SLO. In particular, unlike the RNB case presented in Chapter 4, which has not experienced protests, I also focus on mobilisation strategies of the communities in asserting their rights. I then identifies the findings and possible solutions, and concludes.

The community trust cases in context

The three case studies on which this chapter is based are the Baroka Ba Nkwana Community Trust (Baroka Community Trust) in Atok (Limpopo Province), which is under the leadership of the Baroka ba Nkwana Tribal Authority; the Itumeleng Community Trust (ICT) in Jagersfontein (Free State Province); and the John Taolo Gaetsewe Development Trust (JTG Development Trust) in Kuruman (Northern Cape Province). The majority of the households in the areas of these case studies are living in severe poverty (StatsSA, 2011). I refer to the cases interchangeably by the name of the town or trust.

My first case, Baroka trust, is in Atok, a deep rural village in the Greater Tubatse Local Municipality, which is part of the Sekhukhune District Municipality in Limpopo. Land in Sekhukhune is split between former homeland areas and those that fall outside the Lebowa and KwaNdebele homelands. Land in the former homelands is characterised by low-density residential areas spreading over huge spatial areas. A large proportion of land is under land claims, especially regarding state- and tribal-owned lands, which serve as a barrier for development and future investments (DRDLR, 2016). Greater Tubatse is home to a total population of 490,381 people (Statistics South Africa Community Survey, 2016). The integrated development plan of the municipality indicates that the main constraints to development include a lack of infrastructure (including water, roads, electricity, and sanitation), contestation of land ownership, shortage of skills due to a large pool of unemployed people, and weak institutional capacity to attract and retain skilled staff (DRDLR, 2016). Three mines – Bokoni, Bamanko, and Bauba – operated on five farms under three different tribes. Three of the farms fall within the jurisdiction of the Baroka Ba Nkwana tribe (Baroka tribe), while the remaining two were led by two different tribes. Of particular importance in the case study of Baroka Ba Nkwana Community Trust (BBNCT) has been the relatively long and controversial history of litigation over the chieftaincy between two warring factions of the Baroka tribe in Sekhukhune. The feud is over who should lead the tribe after the death of Chief Nkwana Aubrey Phasha, who died in a car accident in 2003 (*SowetanLive*, 2011). As will be seen, the dysfunctional traditional authority, who are the custodians of the community trust according to customary practice, had a significant influence on the trajectories of the community in seeking an SLO with the local Bokoni platinum mine.

My second case, ICT, is in Jagersfontein, which is a relatively small town in the Kopanong Local Municipality, in Free State Province of South Africa. Kopanong has the largest surface area of the three local municipalities in the Xhariep district, covering 15,190 km^2. Kopanong has immense potential for tourism. The largest dam in South Africa, the Gariep Dam, is situated at the southern border of the region. The potential for tourism of this huge water body is endless. The town has a total population of 1,819, a working-age population of 63.4%, and a dependency ratio of 57.8% (StatsSA, 2011).

Mining began in 1870, and it would yield two of the ten biggest diamonds ever discovered, the Excelsior and Reitz (now known as the Jubilee) before its closure due to the Great Depression in 1930. From 1930 until its final closure, De

Beers Consolidated Mines owned the mine (Davenport, 2011). In 2010, De Beers made good on its stated intention early that year to sell the Jagersfontein mine dumps to a black empowerment firm (Smith, 2010). A few months later, De Beers sold the mine dumps of Jagersfontein Mine to the Superkolong Consortium, a black economic empowerment company, on the basis that Superkolong was the best company that would ensure the community of Jagersfontein benefitted from the mining operation (Seccombe, 2010). The sale agreement included the establishment of a community trust, which was to have a 10% share in Superkolong (Mining Weekly, 2010).

An explicit requirement of the sale agreement was that the Jagersfontein community would be the sole beneficiary of trust funds (Coetzee, 2010a). Interestingly, in what exemplifies the overly broad and indeterminate powers of mining companies locked into the past apartheid mining policy regime in South Africa, Superkolong later concluded a deal with Sonop Diamond Mining and Reinet Investments (listed in Luxembourg) to form a joint venture called "Jagersfontein Developments." However, both the former workers who have been promised a share in the assets of the old mine and a community group – the Jagersfontein Community Trust, which had also applied to mine the Jagersfontein dumps – were excluded from the transaction (Hoo, 2010).

My third community trust case is the John Taolo community trust in Kuruman. The town is in the Ga-Segonyana Local Municipality, the Northern Cape Province of South Africa. The Ga-Segonyana Local Municipality is a former cross-boundary municipality that straddled the borders of the Northern Cape and North West Provinces. The municipal jurisdiction consists of 33 residential areas within a radius of approximately 80 km in and around Kuruman and has approximately 93,000 residents. The municipality's economy is mainly based on surrounding mining and agricultural activities. Kuruman is the municipal seat and central district of the municipality. All the 224 rural villages surrounding Kuruman are administered by traditional authorities (Ga-Segonyana Local Municipality, 2019).

In 2011, the dependency ratio in Ga-Segonyana was 58.1%, and in respect of educational status, 8.4% went through higher education, 29% attended school, and 20.6% attended matric. The unemployment rate was 33.7% in the same year. The economy of the municipality is highly dependent on mining, agriculture, tourism, and the commercial sector in and around Kuruman town. However, the rapid mining developments, while beneficial for the economy, have placed extreme pressure on the ability of the municipality to ensure the availability of water, electricity, waste management, sanitation, and other municipal services (Ga-Segonyana Local Municipality, 2019).

Kuruman also has a long-standing history of violent community protests, including against the then cross-boundary municipality, elections-related and perceived inadequate provision of services by the municipality. The most notable and devastating of these protests have been the so-called Northern Cape "No Road, No School" protests that prevented thousands of learners in more than 50 schools from attending classes, thus costing them a year of learning (Matebesi, 2017).

Atok case – Annooraq Community Participatory Trust

Local concerns and protests

The Atok case is an example of the highly divided and fragmented local group's mobilisation attempts against mining companies and the Department of Mineral Resources (DMR). It also represents a case in which a local pressure group – Baroka Ba Nkwana Community Engagement Committee (BBNCEC), representing 22 villages – has challenged the unilateral decision-making processes between industry and the state about the closure and current plans to reopen the Bokoni mine. The Atok case centres around the Bokoni mine and, to a certain extent, to Samancor platinum mine. Bokoni mine is the oldest of the three mines mentioned earlier. The mine was operated by Anglo American Platinum (Amplats) since the 1960s. In 2009, Toronto-based Atlatsa Resources (Atlatsa) entered into a joint venture with Amplats until it closed the mine in mid-2017. The mine, which has operated under different names and owners for at least 55 years, was placed under care and maintenance in the middle of 2017. This refers to the process of a mine being closed, and production halted, but with the possibility of operations resuming at a later stage. With the closure of the mine, more than 4,500 workers lost their jobs, and business had to close (Ledwaba, 2018).

According to the interviewees, the closure of the mine was, in part, also due to the high levels of labour protests by workers. At times, the workers embarked on strikes that lasted more than five days. As one community leader mentioned, "There is no mining company that would have to survive the pressure that Bokoni was placed under by the workers and community activists." The bone of contention at the time of the closure of the Bokoni mine was that, in 2009, Anglo donated R103 million to Annooraq Community Participatory Trust (Annooraq Trust), which was established to manage the fund. About R24 million was to be invested for the community, while the rest was to buy shares for the community in Atlatsa Bokoni mine. However, for shares to be issued, there had to be a subscription agreement between the shareholding parties. To date, however, Annooraq Trust has not provided the community with updates about its investment nor the subscription agreement. It should be noted that it is not a common practice in local-level agreements that communities are allowed to buy shares in mining companies.

Elsewhere, a colleague and I highlight that the Mining Charter of 2017 may pose several difficulties for the management of community trusts in South Africa (Matebesi & Marais, 2018). In the Atok case, it is not evident why Annooraq Trust was established, while there are several community trusts in the area. Another question that could also be asked in context to the behaviour of Annooraq not demonstrating the inherent pitfalls of SLO is: can communities be disregarded since there is no formal agreement? At a broader level, this is also contrary to collective decision-making by all interest groups, as postulated by the EGT (Beunen et al., 2016).

A series of violent protests were staged against Bokoni mine in November 2014 and February 2015 (James, 2018). The demand was for the Bokoni mine's management to intervene in the dispute of the community with Annooraq Trust. Another episode of violent protest erupted in May 2015. This time employees were prevented from reporting for work due to road blockades and high levels of intimidation. The mine's timber storage yard and network and telephone infrastructure were also damaged. In response, the mine management stated, "No formal demands have been presented to mine management." They continued, "However there was previously an approach to the mine by the Atok Interim Community Structure to engage with Bokoni on Small Medium and Micro Enterprises matters." The management went on to state that Atlatsa proactively engages with business leaders through the Atok Community Engagement Forum, which they claimed was "an independent body elected democratically by community members represented under the three recognised tribal authorities" (Mining Review Africa, 2015).

The literature emphasises that, as an informal process, the SLO is often legitimised by showing a broad favourable public opinion of a mining project, which can be justified and bolstered through the media, for example (Brueckner & Eabrasu, 2018). However, the media and other public relation exercises cannot replace the central role that relationships play in cultivating trust (Fukuyama, 2011; Tilly, 2004; Uslaner, 2018). According to Maddison (2009), there is still much to be done to build Aboriginal trust in some areas in Australia.

By crafting a response that is directed towards business leaders, the management of Bokoni mine lost the opportunity to listen to the grievances of the community. However, BBNCEC maintained that the approach of using protests – save for the most exceptional circumstances – is unsustainable. For example, BBNCEC leaders applied to the Limpopo High Court to issue two declaratory orders. The first declaratory order requested is on the Annooraq Trust about various issues, including the community equity shares. The second declaratory order relates to the legitimacy of BBNCEC as a representative community group in respect of matters relating to the demand of the reopening of the Bokoni mine.

However, BBNCEC soon realised that shifting its repertoires of contention to litigation is a laborious process. Hence, the activist group embarked on a protest to the offices of DMR in Pretoria on 8 November 2018. This was after DMR failed to respond to the petition, with almost 5,000 signatories, submitted to the ministry in March 2018, in which the Atok communities requested the ministry to speed up the regulatory approvals required in terms of the 2017 Restructure Plan of Bokoni mine. In the next section, it will be seen how DMR began a series of consultations. The preceding demonstrates not only the difficult paths to enforce an SLO but also the thorns in the embattled woes of communities to enforce an SLO.

Riofrancos (2019) shows that Latin America – as in many parts of Southern Africa – has experienced an escalation of resource-related conflict in the mining sector. These protests are led by communities directly affected by mining projects. Interestingly, most of the Latin American mining-affected communities are organisationally linked to regional and national indigenous federations, including environmental activists.

Community engagement

The community of Atok began to organise itself in 2014, after it had been ignored for many years. BBNCEC and two other community forums – Roka Selepe and Brakfontein Klipfontein – represented their traditional villages. At the time, the interviewees emphasise, the primary goal was to assess the local impact of mining on communities. The consultation process to enforce an SLO was also complicated by the formation of fake structures. In highlighting his frustration about the fake structures, the deputy secretary of BBNCEC noted,

> I have said this before comrades, and I will say it again. Magoshi only has roles and functions to fulfil, but we the masses, have the power and duty to ensure that we realise our goals with or without the traditional leaders.

BBNCEC further stated that:

> Baroka lets fight against "Hijackers." We had an opportunity to look at Deputy Minister speech on his meeting with Sekhukhune Small Scale Mining Association. One of the minister's suggestions was that the association – in consultation with their chief – must write a letter to the Premier of Limpopo and the Deputy Minister of DMR to indicate that they must be consulted in the application of mining permits. The surprising part is the so-called Teka Traditional Authority. This traditional authority wrote to the deputy minister of DMR and premier. Comrades, it cannot be correct for us to destroy our heritage and culture and work against Baroka Ba Nkwana Traditional House. Teka Tribal Authority is not to be recognised. That is mafia tactics. We are a communal community gazetted as a traditional community under the custodianship of Baroka Ba Nkwana Traditional leadership. People are establishing a new "Moshate now."
>
> [Traditional authority]

This was not the only challenge for the community. The interviewees emphasised that it was initially very difficult to communicate with mining companies. As one noted,

> "We have shares and funds have been allocated for community projects, yet we were not consulted." The consultation process with community structures only began after the protests mentioned before. "However, at the time, we were told that we are fly-by-night structures."

On the one hand, there is widespread dissatisfaction among Atok interviewees with how DMR is dealing with community engagement. On the other hand,

DMR states that the matters raised by communities around Bokoni mine cannot be resolved overnight (Capricorn FM, 2018). As reported above, DMR only started to response favourably after some community members from the Baroka tribe boarded buses to go and protest in front of the DMR headquarters in Pretoria to demand the reopening of the Bokoni mine (Capricorn FM, 2018). According to BBNCEC leaders, four days after the march, DMR sent a delegation of senior officials to meet the Atok community representatives to seek clarity on the memorandum of demands submitted in Pretoria. On 16 November 2018, DMR convened a meeting with Bokoni mine's top management to brief them on the Atok community's primary demands: the community shares subscription agreement and the reopening of Bokoni mine to allow it to fulfil its outstanding social responsibilities.

This meeting between DMR and the mining company was followed by two subsequent meetings with the Atok community delegations in January and February 2019. A major victory for the community is that it eventually received the community shares' subscription agreement, which, according to a post on the Facebook page of Atok Community Development Association, "is in the hands of our technical advisors for further processing." What is worrisome about this Facebook post – also used by BBNCEC to communicate with the community – is the part that "we must say is a confidential document [subscription agreement]." Such a lack of transparency of the information which should be shared with the community suggests two plausible explanations. One of those explanations is that transparency in the torrents of power in the mining industry (government and mining companies) over issues that affect the Atok community is desirable. However, I contend that this principle applies equally, and, even more so, to local mining activist groups with an asserted mission to work for social good. Another possible explanation for the lack of transparency could be the compromising nature of funding. Civil society's primary accountability often shifts to donor practices and, thereby, compromises civic spaces of engagement. In the unlikely event that the Atok community representatives follow the path of a relative lack of transparency and weak accountability mechanisms, they will undermine their credibility among residents.

A second interesting development in respect of the Atok community's attempt to enforce social licensing processes on mines in their community relates to the granting of a chrome-ore mining right to Samancor chrome mine in respect of the farms Wintersveld, Jagdlust, and Zeekoegat. Mining rights were set aside by the High Court in Pretoria on 7 December 2018, due to a lack of proper notice and consultation with interested and affected parties. Interestingly, Samancor Chrome's first application for mining in 2011 was also rejected by DMR for the same reason of a failure to consult. As a result, the properties became open for application, and Bauba, through a subsidiary, lodged an application for a prospecting right over the Zeekoegat property in June 2016. DMR accepted Bauba's application, however, Samancor chrome appealed this decision, which was upheld by the court (Arnoldi, 2018) also due to failure to consult interested or affected parties. BBNCEC maintains that Bauba's application was set aside because it submitted approximately 700 affidavits from local community members, which were questionable. Meanwhile, Samancor has requested leave to appeal, which, at the time of writing (July 2019), had not been heard.

Local impact and benefits

Much of the literature has identified the local benefits of mining as a primary path towards obtaining and maintaining an SLO (Moffat & Zhang, 2014; Moffat et al., 2016). In *Mining Capitalism*, Kirsch (2014) shows the extraordinary history of resource booms and capital deployment in Latin America during the 1990s and 2000s, with extractive projects disrupting the established ways of life. As a result, the region experienced a wave of political mobilisation around mining.

Unlike in many cases reported of unfulfilled promises by mining companies in South Africa (Pickering & Nyapisi, 2017; Corruption Watch, 2018), in Atok, the findings indicate that Bokoni mine financially supported several projects. A total of R50.4 million was allocated by Bokoni mine for CSR projects. In 2013, the Atok community identified 15 projects, which, broadly, related to education, local economic development, and infrastructure development. The interviewees indicated that, in respect of education, two early childhood centres and an administration block at a primary school were built.

Furthermore, the gravel access roads in several villages around the mine were scrapped. An amount of R21 million was set aside to assist the municipality to construct a dam. Other projects included a poultry farm and a skills-development centre. According to the interviewees, only 6 of the 15 projects were completed at the time the Bokoni mine closed its operations. A community leader emphasised that the availability of funds was not an issue in the implementation of identified projects. It all had to do with "the mistake of giving the chiefs the right to decide where these projects should be implemented." A resident also complained about the significant influence of traditional leaders in the processing of identifying land and described that a "lack of water is a huge problem here, yet bakgosi ba rena [our Chiefs] derailed the process of constructing a dam."

Future pathway – deck heavily stacked in favour of community

While most of the leaders representing community organisations are confident that the reopening of Bokoni mine will bring benefits to the impoverished communities of Atok, many residents did not share such sentiments. The leaders indicated that if the government and mining companies work for communities directly during all stages of a mining licence, there is a better chance for improved social and environmental practices. One leader noted that "the regulatory [DMR] used to be a sell-out, but I can see they now understand the importance of allowing us to participate in the process of licensing."

BBNCEC was even more optimistic about the future of mining–community relations in Atok. The group openly declared that the year 2019 "is going to be a turning point of the Atok Economic Revolution." The local group indicated that the days of community representatives driving BMWs bought by mines are over. The vice president further emphasised that one of the main reasons behind the decline in the confidence in mines among community members was the fact that a former DMR senior official became one of the BEE partners of Atlatsa. Disillusioned with such blatantly unethical conduct, the community began to have serious concerns about the credibility of DMR officials. However, it was of the opinion that the situation has since improved, as DMR listens to community concerns.

Some interviewees from the community, however, had little confidence that the reopening of the Bokoni mine will bring any significant changes. They cited the hiring of a hitman to silence vocal community members, and "unholy alliances" between mining companies and some DMR officials. For some, the "revolution is in the hands of the masses," and to others it remains "a cause to die for."

An interesting outcome of the process to enforce an SLO in Atok has led to several reconfigurations in the social fabric of the community. The different local groups have reached an agreement with the local taxi association to ferry residents for free to community meetings held in remote villages.

Jagersfontein case – ICT

Local concerns – a conveyor belt of protests

In a Facebook post on 21 September 2014, a Jagersfontein resident drew the attention of readers to a statement on the website of the ICT, which stated:

> As in other deprived rural areas, many members of the local community have low levels of self-esteem. The Trust envisions engaging with the community to find ways of giving them a new sense of self-worth. This will enable them to function as more responsible citizens and parents, thus enhancing the educational prospects and therefore the life chances of their children.

The resident then proceeded to ask a question: "What is your interpretation or understanding? Is it true that the community suffer from low self-esteem? Let's talk!" The first response to the questions came rather quickly: "I just checked the website, I beg to differ that the community suffers from low-esteem." The response continued, "What I've realised is, we have an angry community that is frustrated due to uncertainty and lack of development." Another response followed:

> If aggressive measures were to be adopted, the community will be well informed about the achievements of the trust. I don't think we would be regarded as low self-esteemed. Information is key, and when you deprive the community of information, how do you expect them to feel? ... *Broer* [brother], the community of Japan [nickname of Jagersfontein], can stand up against such unfounded insults. My view is we are dealing with arrogant individuals of the trust.

As background to the concerns discussed, in another contribution about the Jagersfontein case, Matebesi and Marais (2018) reminded readers that the responses of participants reflected the historical injustices the black population in South Africa have been exposed to. The extract from the website of ICT is a chilling exposé of its corporate value: insensitivity towards vulnerable communities. At the core of this insensitivity, as shall be seen later, is that "later decisions rely on, and are constrained by, earlier decisions" (Hathaway, 2003: 106) or historical ones. Similarly, the subsequent concerns of the residents of Jagersfontein demonstrate not only how the stance of the ICT affected its credibility among the community but also how the trust could not dislodge itself from the historical tangles of the past. Vanclay and Hanna (2019) note that companies need to treat communities with respect and be mindful of local culture as this has a great influence on the chance to obtain and maintain an SLO.

The first set of critical issues identified in my analysis was legitimacy and credibility, two core elements in an SLO process (Thomson & Boutilier, 2011; Edwards et al., 2018). According to IGOPP (2018), "a board's credibility is a condition *sine qua non* of its effectiveness and ability to add value" for improving the performance of an organisation for all its stakeholders. A major complaint from the participants in my study was that a board of trustees was elected in 2011 when ICT was formed. Since then, no public elections have been held. There is also a fallout between the trustees who are residents of Jagersfontein and the local activist group, Jagersfontein Multi-Stakeholder Mining Forum (JMMF). The analysis further revealed that JMMF's concerns about the legitimacy of the ICT board were about who nominate members for the board and how are they elected.

Another major concern of JMMF and interviewees was that board members "are quick to state that they cannot say anything about the board as they signed a confidentially agreement" when asked about their role. Several interviewees believe that the reason for the intransigent attitude of local board members could be ascribed to the R5,000 monthly and R20,000 annual allowances they receive, and the "end-year flights to go and spent some time in Cape Town." This was, however, refuted by ICT's secretary in an interview. This did not, however, stop the community from embarking on several protests against ICT over a lack of transparency, accountability, and its approach to local development. These protests – signalling the absence of an SLO – began in May 2011, when community representatives demanded 800 jobs (Boucher, 2012). Subsequent protest marches were held on 10 February 2015, 14 January 2016, and 23 May 2017 (Matebesi & Marais, 2018).

Another march was held on 22 November 2018. This time the community was supported by residents from nearby Fauresmith, a small town 11 km from Jagersfontein, and a few provincial members of the South African Communist Party. This time the main demand was for ICT to inform the community about the R20 million donated by the Rupert Foundation for CSR projects (Coetzee, 2010b) and the ongoing lack of authentic engagement with the local community.

In what discerns the growing support for the Jagersfontein community, Motse (2019) reported that on 5 December 2018, the local municipality intervened by shutting off the supply of water to the mine to force it to discuss the issues raised by the community. However, on 13 December 2018, the mine turned to the High Court to obtain a court interdict to prevent, the municipality and JMMF from trying to prevent repair work that the mine wanted to be done to the water infrastructure belonging to the company (*OFM News*, 2019).

The pathways of conflict continued in 2019 when a leader of JMMF wrote an open letter to DMR and the Premier of the Free State Province, requesting assistance in the community's attempt to interdict the operation of Jagersfontein Developments. The first reason cited in the open letter was that the mining company was digging virgin ground without a permit. The digging, the open letter notes, takes place next to the open cast hole, and this is a threat to the community. A request was further made to DMR to protect the rights of the community. The open letter ends with a clarion call by the author: "We have had enough of being treated like we don't exist and the audacity by one of the whites to say in our faces that as long as the ANC is in power, we won't shake them."

The reference to race is a path-dependence response keeping in line with the tradition of using the race card whenever there are differences of opinion in South Africa.

Community engagement

The Jagersfontein case has, until now, highlighted the challenges of constructing an SLO, as well as how the interests of the mining company prevailed each time there was conflict. However, the management of ICT denied that there is no authentic engagement with the community. ICT mentioned that attempts to engage the community of Jagersfontein through public meetings failed as the focus was on problems. Disillusioned with the disruptive and, at times, abusive behaviour of some community members, ICT decided to seek alternative means of community engagement: quarterly board meetings as well as newsletters. When asked why community relations with ICT were weak, the ICT official interviewed said, "It is only a few individuals who are dissatisfied. It is always the same group of people who are complaining."

Scholars have argued that achieving community trust through direct involvement and representation depends not only on the engagement process but also on the legacy of past activities and relationships (Cornwall, 2004). Several documents indicated that ICT conducted a need analysis in which the community participated and identified several projects, including a stadium. The issue of the stadium was mentioned by several residents as one of the initial reasons why the community began to question whether the engagement by ICT is genuine. The interviewees complained that the trust tasked a few people to look into the design of the new stadium, only to reject the project at a later stage. This argument by the community, it can be contended, could have had deleterious consequences for the ICT and the community as a stadium is not an economically sustainable project. Furthermore, the arguments demonstrate path dependence locked into a history of unrealistic expectations of the role of mining companies.

Local impact and benefits

This section examines the local impact and benefits of mining in Jagersfontein from the perspective of both ICT and community members. The stated mission of ICT (n.d.), as stated on its website, is to develop and empower the communities of Jagersfontein, Itumeleng, Charlesville, and the surrounding farming areas, using public benefit activities and interventions in an independent manner and through good corporate governance. ICT believes that it has made a meaningful impact on the lives of locals. An assessment of ICT's website shows that many of the projects are funded by ICT, in collaboration with the Rupert Foundation (ICT, n.d.).

These projects include the provision of bursaries for 12–14 qualifying students to study at the South African College for Tourism in Graaff-Reinet, in the Northern Cape Province. The male students are taught indigenous tracking skills and general hospitality at the Samara and Londolozi game reserves, while the female students study hospitality. According to the ICT newsletter (ICT, 2015), its recycling project has contributed R570,000 a year to the town's economy and employed more than 180 people since its startup in the second quarter of 2012. A major spinoff of the recycling project is that more than 90 tons of waste have been removed, making a significant contribution to the cleanliness of the town. The newsletter further reports that the yearly food parcel project – funded to the tune of R150,000 by 2015 – has a total of 300 recipients (ICT, 2015).

ICT also manages a skills training programme, a community recycling project, a food parcel distribution project, three educational projects, and several sports initiatives. It has also established a small project fund, which enables local trustees to support local organisations and initiatives rapidly. ICT is also working to strengthen ECD in the target area. Wooden playground equipment has been donated to an intermediate school for its kindergarten classes. More than 70 learners are benefitting from this first-class equipment. Plans are in the pipeline to provide similar equipment to all the crèches in the target area. The ICT was responsible for the initial negotiations with Checkers to open a USave Supermarket in Jagersfontein, which will help to reduce the cost of living for local and farming communities (ICT, 2015). This supermarket has a huge impact on household expenditure on food. Previously, some residents of Jagersfontein and surrounding towns travelled to Bloemfontein – 110 km away – to do their monthly grocery shopping.

An important reason communities care about the fairness of enacted procedures for decision-making about representation and engagement is to ensure progress towards sustainability through benefits (Veiga et al., 2009). As noted before, residents who view the board of ICT as unfairly constituted, or generally have a negative view of the trust, are naturally predisposed to label the local benefits of mining negatively. My findings reveal that few interviewees indicated that they are satisfied with the job opportunities, food parcels, and bursaries offered by ICT.

However, the majority of the interviewees mentioned viewed the projects undertaken by ICT as "handouts," "inadequate," "condescending," and "disrespectful," and seemingly bearing no relation to the "original projects the community had agreed upon." One local business owner stated that "the so-called projects and donations are part of a well-orchestrated plan to legitimate an illegal institution." JMMF leadership reported that they are concerned that the projects of ICT bear no relation to the sustainable projects which the community wanted to have implemented. They declared that most of the projects are exactly the opposite of what the impoverished community needed. As one remarked, "While we appreciate that the bursaries can go a long way in liberating the youth in this town, who funds these bursaries?" According to the Jagersfontein Multi-Stakeholder Mining Forum (JMMF), most of the projects and programmes of ICT are funded by the Rupert Foundation and could therefore not understand what happened to the R20 million. JMMF was convinced that ICT seems to pursue short-term projects at the expense of long-term sustainable projects.

My findings with regard to the local benefits of mining are consistent with those of a host of studies that mining-affected communities benefit relatively less from mining projects. However, this finding should be understood within the context that in the current operations of Jagersfontein Developments, tailing is not subjected to the provisions of the MPRDA in respect of community benefits.

Future path – renewed quagmire of contestation

Until now, I have been concerned with how the troubled history of the social licensing process in Jagersfontein has unfolded. Now, the focus turns to the possible future paths in the endeavours to enforce an SLO in Jagersfontein. Research in more than 22 case studies of community protests in South Africa (Matebesi, 2017; Matebesi & Botes, 2017) has shown how unresolved community grievances create a vicious cycle of protest episodes in affected communities. Available evidence points to a similar situation in Jagersfontein. This time, however, the target of community mobilisation will include its leaders. How did I arrive at this conclusion?

The manager of ICT stated, in the interview with me – and confirmed by Motse (2019) – that the main leader of the JMMF was employed for approximately five months by ICT. He was later dismissed after a disciplinary hearing. After his dismissal, the leader of JMMF made several requests to ICT for the funding of various projects and business ventures. However, ICT declined the requests as they fell outside the mandate of the trust (Motse, 2019). This, the manager of ICT believes, provoked the wrath of the leader "whose frustrating and continued criticism of the Trust is unwarranted."

However, JMMF declared they have no confidence in ICT. In early 2019, the same leader of JMMF, circulated a draft letter requesting the assistance of the Free State Legal Practice Council. The letter was calling for assistance from the council to represent him "on behalf of the community to pursue this matter regarding the complete implementation of the De Beers Sale Agreement and the nullification of the Trust Deed in its current form." Three months later the developments in Jagersfontein situation took a drastic turn. Jagersfontein Developments, which until recently, have been processing the mine tailings, applied for a prospecting licence for full-scale mining. As a legal requirement, public consent has to be obtained. Meanwhile, JMMF showed its displeasure with the new unfolding development by writing a letter to DMR in which they categorically stated they are against the issuing of a prospecting licence to a company that has been disrespecting the community. Interestingly, but not surprising, the two leaders of JMMF claimed to have been offered an opportunity to act as intermediaries – for a monthly fee – between the mining company and the community.

From a community perspective, there are measured but sobering warnings that this kind of secret dealings – if true – will have a profound and long-lasting effect on the Jagersfontein community's capacity to enforce an SLO. This is particularly the case in a country where society's confidence in transparency, accountability, and fairness has plummeted (Kingston, 2018). Whatever happens over the next few months between the mining company and the community, the company will continue to find itself between a rock and a hard place with respect to its credibility. Research has shown that decision-making procedures enacted fairly promote cooperative behaviour (Veiga et al., 2009). Jagersfontein Developments, JMMF, and other interested stakeholders and the community of Jagersfontein have a real opportunity to make a complete break with the past. It is clearly beyond doubt – after all – that the ICT-community relations are on a path of a renewed quagmire that is firmly rooted in the ten-year history of the mine project.

Kuruman case – JTG Development Trust

The JTG Development Trust was established to improve the quality of life of the people of the JTG District Municipality in the Northern Cape Province. The trust is named after the district municipality, which – in recognition for the contribution of long-standing and respected member of the African National Congress JTG in the struggle against Apartheid – was renamed the Kgalagadi District Municipality in 2008. JTG Development Trust – hailed as a success of a truly transformed South African mining investment company – was previously known as the Kgalagadi Rural Poverty Node Charitable Trust (*Solomon Star*, 2018).

JTG Development Trust receives funding from the Sioc Community Development Trust (Sioc CDT), popularly known as the "Super Trust." The Super Trust was officially launched in December 2008. Kumba Iron Ore (Kumba) – which owns iron ore mines in Thabazimbi (Limpopo) and Sishen (Northern Cape) – is the majority shareholder in the Sioc CDT. The latter has four beneficiary trusts, including JTG Development Trust (*City Press*, 2013), which serve as implementing agents for community projects.

The trust is also one of the 16 shareholders in Ntsimbintle, a black economic empowerment manganese mining and exploration company. With a 14.4% shareholding stake in Ntsimbintle, JTG Development Trust has received dividend payments amounting to R46.2 million from Ntsimbintle by 2017. This has greatly empowered the trust to carry out its mandate to champion sustainable socio-economic solutions for the poor and needy people of the Kgalagadi district (*MiningNews*, 2017). The next section focuses on the local concerns of Kuruman residents regarding the Kuruman branch of the JTG Development Trust.

Local concerns in Kuruman

In *Subterranean Struggles*, Bebbington and Bury (2013) provide a comprehensive examination of extractive industries. The scholars analyse the dynamics of the subsoil and illustrate how social mobilisation and resistance, for example, are engendered by pressures exerted by resource development. The political ecology of the subsoil is important for understanding the dynamics of power and governance of extraction.

As noted earlier, academic contributions on CDTs in South Africa are relatively scant. JTG Development Trust staged its meetings under a tree upon being founded in 2002, but today it boasts comfortable offices from where it renders support to the three local municipalities: Joe Morolong, Gamagara, and Gasegonyane. The CEO of the trust, it is reported, has been forthright on accountability: "The Trust prides itself with good governance. We are audited by an independent and reputable auditing firm and have achieved successive clean audits" (*Solomon Star*, 2018).

However, for the community leaders of Kuruman, JTG Development Trust has failed for almost ten years – at the time of the fieldwork in 2017 – since it was established. The four community leaders interviewed categorically said they welcome the digitalisation of communication. However, they believed that it diminishes public interest in the trust. The leaders emphasised that they have sought every possible recourse, including the writing of emails, letters, and even personal visits, to meet the management of the trust, but found little solace in replies.

As a last resort, they wrote to the internal fraud units of two mines – Kumba and Sishen – requesting their intervention. The leaders complained that the community was in the dark about the trust's financial performance; there was no communication between the trustees and the community, and many projects started by the Trust are not completed. As a *City Press* (2013) report also noted, the leaders complained about the rampant corruption, as most of the projects the trust initiates are being implemented by companies with a close connection to the board of trustees. Subsequently, the Kimberley Commercial Crime Investigation Unit began an investigation into the affairs of the trust (*City Press*, 2013).

However, and surprisingly so, the spokesperson of the Super Trust at the time said there was no sinister motive for the timing of the annual general meeting. He said that the JTG Development Trust was adopted as a beneficiary trust during 2006 by the Super Trust, and, between 2006 and 2011, the trust received funding to strengthen the operations and to kick-start community initiatives (*City Press*, 2013).

While the lack of control over the decisions of the trust was heightening, the leaders took a more activist stance by mobilising the community in an intensive public awareness campaign. This campaign was to make the community aware of the environmental, health, and other social impacts of mining. Above all, pointed out one leader, "it was the calculated plan by some trustees to discredit and intimidate us." The leaders were driven by a sense of responsibility for ensuring – knowing that mining will not continue indefinitely – that the little the community receives from mining is effectively managed. "If the mines did not listen to us, we were prepared to shut them down," said one leader. Another one said the community was unanimous in its desire to see the community continue to have the right to provide informed consent to the operations in mines, and represent and participate in their local trusts.

Another concern raised by the leaders, as well as some of the residents interviewed, was the lack of transparency in the election of the board of trustees of JTG Development Trust. Some residents pointed out that the board of trustees has been highjacked by the local political elite. Overall, a few residents were satisfied with the performance of the trust, but the rest believed that it seems not to be prepared to embrace arrangements that support inclusivity. This way, the interviewees noted, was a systematic undermining of the community.

Young respondents were generally distraught mainly by their state of unemployment. They felt that they should have been in the frontline of the queue when it comes to employment and training opportunities provided by local mining companies. The issue of employment by local mines, it must be stated, was one of the causes of the intermittent, yet devastating and destructive, community protests that began in Olifantshoek – a small town, 60 km away – and spread to Kuruman between 2012 and 2016 (Matebesi, 2018). As reported earlier, the protests led to the closure of schools for almost a year and, I believe, placed pressure on local municipalities and the Northern Cape provincial government.

Within such a volatile environment, I contend, the JTG Development Trust could no longer ignore the request of the community. However, by the time trust acceded to the community's request for a meeting, it was a little too late, as it was going to be difficult to shake off the view of an organisation that has been cloaked in a governance tangle of indifference towards the community of Kuruman. This point is in close agreement with the main assumption of this study that institutional lock-in that occurs in CDTs is a result of historical practices dealing with, particularly, historically disadvantaged communities in South Africa.

Community engagement

Generally, protest actions – in their various forms as illustrated in the preceding section – are considered the expression of grievances, which could be addressed by engaging in meaningful dialogue with protesters (Vanclay & Hanna, 2019). The point to be illustrated here concerns the complaints of the community leaders of Kuruman, which could have been used as an opportunity to solve problems and enhance the adversarial relationship between the trust and the community.

However, my interviews with the leaders of the community reveal the opposite occurred. At the annual general meeting of JTG Development Trust, the leaders soon realised that they were unwelcome. The chairperson opened the meeting by stating that the purpose of the meeting is to deal with the five items on the agenda, and under no circumstances will disruptive behaviour be allowed. The last item on the agenda, the leaders continued, dealt with the audited financial statements for 2011. The statement indicated that the JTG Development Trust received R102 million, of which R65 million was spent on special projects. One of the community leaders wanted to know what these special projects entailed. According to the leaders, no answer was provided and the meeting was adjourned abruptly. According to *City Press* (2013), the Super Trust 2011 annual report shows that R102.6 million had been approved for 15 projects in the 2011 financial year.

In my interview with the Project Manager of the trust in March 2016, he confirmed that the annual general meeting took place, but denied that they had neglected community engagement. He said that the trust could try to be more responsive, visible, and available for engagement and consultation. It should be stated, since the fieldwork, JTG Development Trust has made significant strides to enhance community engagement. Thus, community mobilisation has managed to disrupt the traditional paths of community engagement that avoid tokenism.

Local impacts and benefits

One notable feature related to the local impact of mining is the structuring conditions, like the racial segregation policies that led to separate racial development trajectories in South Africa, for example, that persisted for a long time and further condition other historical developments. To meaningfully reform those systems requires a disruption of the institutionalised structures that support inequity or what Mahoney (2000: 511) referred to as "the formation and long-term reproduction of a given institutional pattern."

There is also the unrealistic expectation that mining will resolve all the historical challenges a mining-affected community experiences. Marais et al. (2018), for example, highlight the difficulties that mining towns experience in dealing with both a boom and a bust cycle. As a result, mining towns are locked into pathways of perpetual conflict, and, more so, in local communities' efforts to enforce an SLO (Matebesi & Marais, 2018).

In this study, it is evident that the JTG Development Trust, despite the intense criticism about corruption and a lack of community engagement, constitutes one of the rare examples of CDTs that have meaningfully contributed to the well-being of a local community. The trust has established a successful bursary scheme, constructed a community hall and a clinic, provided learning institutions with scientific equipment, equipped computer centres, and upgraded sporting facilities (*Solomon Star*, 2018). By 2018, a total of 237 bursary recipients had completed their university studies. Furthermore, the trust has been offering free computer training for end-user computing, and 1,228 have been found competent since 2011 (NCNN, 2018).

In respect of health care, the trust's *Re A Fola* Health Promotion project has brought health-care services closer to communities in the form of health-care facilities and mobile clinics. This project includes a health promotion workshop, which focuses on several communicable and terminal diseases and aspects of a healthy lifestyle, as well as provides access to a general practitioner, a dentist, and an optometrist. The success of this project prompted other community trusts, such as the Gamagara Development Forum, to adopt a similar approach (Gamagara Development Trust, 2013). In addition, the Northern Cape Provincial Department of Health entered into a partnership with the JTG Developmental Trust for the procurement of medical equipment for hospitals, CHCs, and clinics in 2012. The total cost value of these projects exceeds R30 million (Office of the Premier-NC, 2012).

The trust has also been involved in projects of infrastructure development, such as the building of schools, clinics, housing, community centres like halls, and tribal authority offices, among many other initiatives it had undertaken (NCNN, 2018). These benefits will create a path dependency on mining, as is the case with many northern communities (Bennett, 2016; Petrov, 2016).

The way forward for the JTG Development Trust

There are several unresolved challenges that JTG Development will need to address in future. Overall, the interviewees had no confidence in the trust's ability to change their lives in future. Their responses were based on the lack of trust in the credibility of trustees and the management of the JTG Development Trust. The trust's responsibilities cover a large area, and the widely perceived selective targeting of certain neighbourhoods over others has contributed to an increased lack of trust and credibility – key elements in the paths to an SLO (Moffat & Zhang, 2014) – among the residents of Kuruman. This lack of trust is further observed among the younger participants who believe that the JTG Development Trust is incapable of addressing the challenges of the youth in Kuruman effectively.

As seen in the preceding sections, the major risk – and opportunity – for the JTG Development Trust is community engagement. Despite the many development projects the trust has managed to initiate successfully, the lack of trust and credibility in the ability of the community to participate in the decision-making processes of the trust will continue to be a major area of conflict. The community will seek to assert its right to enforce an SLO, which, if not granted, may lead to mobilisation against the trust and mining companies. Again, the trust should understand that it cannot act as if its operations are so obvious that it does not require much communication. The trust will be best suited to looking at other mining-affected communities in South Africa and beyond for an illustration of the significance of community engagement on social acceptance.

Perhaps, most importantly, creating real benefits for communities will be key for the way forward for the trust. The trust also has to deal with the perception that it has become a repository for illicit business transactions by trustees and management. A Mpumalanga provincial manager of Mining Affected Communities United in Action South Africa (MACUA-SA) (interview 14 November 2017) confirmed that the misleading of communities by a politician who has business interests is a nationwide challenge for CDTs. These trusts serve as the mechanism by which political influence reigns over corporate good governance practices.

It is not clear if the trend of disregarding the community will continue, but what is certain, as expressed by the interviewees, however, is the need to ensure that the wheels of systemic greed characterise local community-based resource management institutions to arrest the path of cynicism and despair among mining-affected communities. These challenges are not insurmountable for the current capable leadership of the JTG Development Trust.

Survey cross-comparison of the three case studies

This section provides a brief comparison of the survey results of the three case-study residents. The focus is on procedural fairness and trust in the CDT, perceptions about the contribution of the CDT to the quality of life, and future behavioural intentions. In respect of gender and age, the Kuruman case reported more males than the other two cases. The Jagersfontein case had younger respondents, followed by the Atok case.

Of the 200 respondents to the Atok survey, the majority were males (59%, $n = 118$) and 41% ($n = 82$) females. The largest percentage of respondents were in the age categories of 21–35 years (44%) and 36–55 years (47.5%). In respect of educational level, the majority (58.2%) had completed primary school, while 10.5% had no formal schooling at the time of the survey.

The survey of the Jagersfontein residents shows that slightly more female respondents (53.5%, $n = 107$) than males (46.5%, $n = 93$) participated in the survey. Half of the respondents were in the age category of 36–55 years (51%) and a third in the age category of 21–35 years (32.5%), while 16.5% were 55 years and older. Slightly more than half of the Jagersfontein survey respondents (67%) and slightly more than a third (34.6%) had completed high school and primary school, respectively. About 9% had obtained a diploma or degree.

The respondents' demographic background in the Kuruman survey shows a skewed gender ratio in favour of male respondents, at 64% ($n = 128$). Respondents in the age category of 21–35 years were 37%, while the most were in the middle-age category of 36–55 years (54.5%). In respect of educational level, the majority had completed primary school (61.5%) and 21.5% high school.

Overall, Table 5.1 shows that more survey respondents from Kuruman (37.7%) and Jagersfontein (33.3%) reported that they believe that the CDT in their town listens to and respects their opinions. The Kuruman case had more respondents (48.6%) reporting that they believe that their CDT is prepared to change its practices in response to community sentiment, compared to Atok (25.1%) and Jagersfontein (36.9%).

130 Social mobilisation

Table 5.1 Procedural fairness and trust in CDT (percentage in parentheses)

Variable	Atok	JFN[A]	Kuruman
	Yes		
Procedural fairness			
I believe the CDT listens to and respects our opinions	28 (13.8)	67 (33.3)	75 (37.7)
I believe the CDT is prepared to change its practices in response to community sentiment	50 (25.1)	74 (36.9)	97 (48.6)
Trust in CDT			
I can trust the CDT to act responsibly	35 (17.4)	56 (28.0)	71 (35.3)
I trust the CDT to act in the best interest of the community	43 (21.5)	78 (38.9)	85 (45.7)
My trust in the CDT has decreased over the past five years	173 (86.8)	153 (76.6)	112 (56.0)

A JFN = Jagersfontein.

In respect of the trust, the Kuruman respondents are more frequently able to trust their CDT to act responsibly (35.3%), while the Atok respondent had the least trust that their CDT will act responsibly (17.4%). Atok also had the least reported respondents who trust their CDT to act in the best interests of the community, but by far, the most respondents who indicated their trust in their CDT have decreased over the past five years (86.8%) compared to the Jagersfontein and Kuruman cases.

In stark contrast to the two other case studies, Kuruman survey respondents overall eclipse the respondents in the other two cases in respect of believing that their CDT makes a major contribution to the quality of life of residents (Table 5.2). A closer look reveals that 46% and 40% of female respondents in Atok and Kuruman, respectively, believe that their CDT makes a major contribution to the quality of life of residents. In the case of the former, 43% of older respondents and 81% of those who obtained primary schooling also indicated that they believe that their CDT makes a major contribution to the quality of life of residents.

The respondents in Atok are by far more likely to act affirmatively in respect of behavioural intentions. For example, 88% of the survey respondents from the Atok case reported they would likely complain to people about their CDT, followed by 79% in Jagersfontein. More respondents from the latter case reported being likely to attempt to discourage others from working with their CDT (61%), while most Atok respondents (62%) indicated that they are likely to participate in a protest by residents against their CDT compared to those in Jagersfontein and Kuruman. There was no major difference among respondents in Atok and Jagersfontein, who said it is likely that their community will be a better place to live in a decade from now (Table 5.3).

Table 5.2 CDT's contribution to the quality of life of residents (percentage in parentheses)

Variable	Atok	JFN[a]	Kuruman
	Yes		
Gender			
Male	19 (16.4)	28 (26.1)	33 (25.8)
Female	38 (46.0)	17 (18.3)	29 (40.3)
Age			
21–35	8 (8.7)	24 (23.5)	26 (35.1)
36–55	23 (24.2)	14 (21.5)	43 (39.4)
56 and older	7 (43.7)	9 (27.2)	5 (29.4)
Educational qualification			
No formal schooling	12 (44.4)	2 (33.3)	9 (40.9)
Primary school (Grades 1–7)	94 (81.2)	13 (26.0)	56 (45.5)
High school (Grades 8–12)	8 (17.0)	12 (9.6)	7 (16.2)
Postgraduate	0 (00.0)	2 (11.1)	1 (8.3)

a JFN = Jagersfontein.

Table 5.3 Future behavioural intention regarding selected variables (percentage in parentheses)

Variable[a]	Atok	JFN[a]	Kuruman
	Likely		
How likely will you complain to people about the CDT?	176 (88.1)	158 (78.9)	135 (67.4)
How likely will you attempt to discourage others from working with CDT?	99 (49.3)	122 (61.2)	83 (41.7)
How likely will you encourage others to protest against the CDT?	153 (76.7)	105 (52.4)	72 (35.9)
How likely will you participate in a protest by residents against the CDT?	124 (62.2)	82 (41.0)	55 (27.5)
How likely will your community be a better place to live in a decade from now?	30 (15.0)	24 (12.0)	49 (24.5)

a Measured with a seven-point Likert scale.

The survey results provided indications of the perceptions of respondents in respect of procedural fairness and trust, as well as the quality of life and behavioural intention. I now turn to a discussion of the study findings.

Discussion – emerging paths to SLO in South Africa

This chapter examines the process of achieving locally based social licences through CDTs in three locations in South Africa. Through the useful lenses of path dependency within EGT, the chapter further illuminates the book's central thesis that mining areas are locked into pathways of conflict that persist irrespective of the institutional arrangements that have been devised to address this reality. Theoretically, there are diverse views on SLOs (Owen & Kemp, 2013). The broad question underlying the diverse views is how the licence is granted in reality. The basic tenet is that SLO "is a theoretical construct representing the implied consent by affected stakeholders for businesses to operate, independent from legal or statutory requirements" (Taylor & Mahlangu, 2017) and that SLO is used in extractive industries both as "a response to calls for greater community engagement and as a corporate sustainability strategy" (Meesters & Behagel, 2017: 274).

The cases studies presented in this chapter have experienced a high incidence of protests, though this varies considerably in respect of intensity. Bebbington and Bury (2013) offer insight into the territorial dynamics and political conflicts associated with social licensing in Bolivia, Peru, and Ecuador. Using a political ecology perspective, the scholars demonstrate how extraction intersects with economic and political processes, as well as how these processes are shaped by historical and geographical contexts. Following the work of Bunker (1989) and Boyd et al. (2001), Bebbington and Bury (2013) position mining as a form of land use, source of livelihood, and elite political project that contribute to conflict in mining-affected communities.

In this study, there is evidence of a huge void in the democratic election of the board of trustees of CDTs. These trusts, the cases demonstrate, represent a toxic mix of maladministration, while local activist groups face serious challenges in addressing community engagement processes that are often delivered in jest.

In reality, a significant part of the response of industry and the government to mining-affected communities is based more on history than anything else. In the context of South Africa, that history has been punctuated with the social and economic exclusion of the majority of the black population. For example, it seems DMR is locked into past practices despite its mission to be a leader in the transformation of South Africa through economic growth and sustainable development, by 2030. The clandestine nature in which the department deals with mining-affected community concerns has contributed to the conflict in these areas. It is therefore not an overstatement to conclude that the department resembles an institution that is cloaked in a legislative tangle. For the sake of vulnerable communities, DMR needs to deeply consider the decisions it makes and examine if it is in line with the department's ostensible objective to ensure local ownership of natural resources in South Africa.

Dare et al. (2014) connected the issue of community engagement with its unique role in achieving social licences at multiple scales. These scholars identified three key challenges that inhibit operational community engagement processes from achieving various social licences: issues of trust, representation, and changing expectations. They conclude that an SLO is achieved through the cumulative attainment of social legitimacy, management credibility, and trust "achieved primarily through communications associated with effective and genuinely conducted community engagement activities" (Dare et al., 2014: 195). Furthermore, as shown in Joyce and Thompson (2000) and Bull et al. (2008), successful engagement experiences create positive reputational capital, which influences the development of trust, credibility, and, consequently, the SLO.

However, residents in all three cases bemoan the lack of accountability and involvement in the decision-making processes of the CDTs. To illustrate, consider the cases of Kuruman and Atok. In Kuruman, the JTG Development Trust failed to consult the community for more than ten years, while the Annooraq Trust has not been sharing any information regarding the subscription shares of the community. Similarly, the community of Jagersfontein has been short-changed, in numerous ways, by the local municipality, provincial government, and activists themselves. This situation presents a huge barrier for mining communities to break away from the entrenched path, reinforcing the ubiquitous corruption that pervades every level of South African society.

The survey results also reveal an overall lack of confidence in CDTs and their ability to improve the lives of residents in the next decade. While a cursory view of the survey results shows that Atok residents have the least trust in their CDT, the overall results are in line with the study of Walsch et al. (2017), who found that the community experience of consultation drove negative perceptions of the proposed mine due to procedural factors. Similarly, utilising a path-dependent framework, Jaskoski's (2014: 873) study on Peru's mining sector found that "very limited spaces for community participation in the environmental impact assessment process, in fact, have prompted and transformed popular mobilisation in extractive zones."

A study in New Zealand (Edwards et al., 2018, though preliminary, identifies the contribution of trustworthiness to the establishment of trust. The scholars believe that "a re-introduction of face-to-face and more personal contact is necessary for trustworthiness and trust due to the importance of contact quality" (Edwards et al, 2018: 8).

This study reveals the remarkable extent to which local communities respond to the increasingly condescending posture of CDT management and mining companies. These CDTs, the study reveals, have become the bedrock of social mobilisation by grassroots activists for transparency and accountability (Pickering & Nyapisi, 2017) and the management of mining royalties in South Africa (Corruption Watch, 2018). These events have inherently created a system of differential risks and opportunity-threat impact for local stakeholder groups and residents. While social licensing can play a significant role in bridging the gulf between industry, government, and society, the paths to achieving an SLO seem daunting. However, in line with EGT theory, which notes the significance of community-inclusive processes that build trust and confidence in actors and institutions, one major lesson is evident: as frustrating as it may be, local communities do not easily give up in their quest to enforce an SLO.

Conclusion – a robust project-level recourse to SLO

At present, the one-size-fits-all approach to obtain, maintain, and grant an SLO requires differentiation according to the local contexts of mining-affected communities. A clear distinction should be made, for instance, between the implementation of CSR projects and community engagement. Because of the high levels of subtle intimidation by mining companies, CDT leaders, and local elites, some community members are discouraged from taking an activist approach to conflict over community-based mining management practices. This effective threat to the communities not only undermines their ability to exert influence but circumvents community consent.

The paths to SLO have to recognise two fundamental factors. First, it has to recognise the systemic and institutional barriers for mining-affected communities to enact an SLO. Second, ensuring a robust project-level-recourse to an SLO regime that effectively addresses legitimacy and trust is essential to changing the current untenable paths of community-based resource management that are locked into past legislation and amplified by a democratic state. Consequently, the analysis here provides all actors within the complex nexus of SLO dynamics an opportunity to critically reflect on their roles in mining company–community relations. However, CDTs across the world, in particular, would be remiss if they did not at least embark on developing institutional design principles to address participatory governance mechanisms for mining revenues. In this way, the stakeholders collectively will be able to disrupt the pathways that conflict-mining–affected communities are locked into.

References

Arce, M. (2014). *Resource extraction and protest in Peru*. Pittsburgh: University of Pittsburgh Press.
Arnoldi, M. (2018). Bauba says High Court halts mining right grant to Samancor Chrome [online]. Available at www.engineeringnews.co.za/article/bauba-says-high-court-halts-mining-right-grant-to-samancor-chrome-2018-12-07 (accessed 19 July 2019).
Bebbington, A. & Bury, J. (eds). (2013). *Subterranean struggles: New dynamics of mining, oil, and gas in Latin America*. Austin: University of Texas Press.
Bennett, M.M. (2016). Discursive, material, vertical, and extensive dimensions of post-Cold War Arctic resource extraction. *Polar Geography*, 39(4): 258–273.
Beunen, R., Van Assche, K. & Duineveld, M. (2016). *Evolutionary governance theory*. New York: Springer.
Bice, S. (2014). What gives you a social licence? An exploration of the social licence to operate in the Australian mining industry. *Resources*, 3(1): 62–80.
Boucher, E. (2012). Corporate social investment: A tale of two towns. *Vrystaat Sake Bulletin*, p. 1.
Boyd, W., Pruham, W.S. & Schurman, R.A. (2001). Industrial dynamics and the problem of nature. *Society and Natural Resources*, 14(7): 555–570.
Brueckner, M. & Eabrasu, M. (2018). Pinning down the social license to operate (SLO): The problem of normative complexity. *Resources Policy*, 59(2018): 217–226.
Bull, R., Petts, J. & Evans, J. (2008). Social learning from public engagement: Dreaming the impossible? *Journal of Environmental Planning Management*, 51(5): 701–716.
Bunker, S. (1989). Staples, links, and poles in the construction of regional development theories. *Sociological Forum*, 4(4): 589–610.
Capricorn FM (2018). Demand to open Bokoni platinum mine [online]. Available at www.capricornfm.co.za/index.php/2018/11/09/demand-to-reopen-bokoni-platinum-mine/ (accessed 24 July 2019).

Centre for Applied Legal Studies (2017). *The social and labour plan series phase 2: Implementation operation analysis report.* Johannesburg: University of Witwatersrand.

CERLAC (Centre for Research on Latin America and the Caribbean) (2002). *Canadian mining companies in Latin America: Community rights and corporate responsibility.* A Conference organized by the Centre for Research on Latin America and the Caribbean at York University. CERLAC and MiningWatch Canada, May 9–11, 2002. Toronto, Canada [online]. Available at https://miningwatch.ca/sites/default/files/CERLAC_mining_report.pdf (accessed 14 June 2018).

City Press (2013). Sishen community trust under investigation [online]. Available at www.news24.com/Archives/City-Press/Sishen-community-trust-faces-fraud-probe-20150429 (accessed 11 June 2016).

Coetzee, E. (2010a). Woede, oproer maak plek vir hoop. *Volksblad*, p. 1.

Coetzee, E. (2010b). Woede, oproer maak plek vir hoop op dorp. *Volksblad*, p. 5.

Community Survey (2016) [online]. Available at www.statssa.gov.za/?page_id=993&id=greater-tubatse-municipality (accessed 12 September 2017).

Cornwall, A. (2004). Spaces for transformation? Reflections on issues of power and difference in participation in development. In: S. Hickey & G. Mohan (eds). *Participation: From tyranny to transformation? Exploring new approaches to participation in development* (75–91). London: Zed Books.

Corruption Watch (2018). Mining royalties report 2018 [online]. Available at www.corruptionwatch.org.za/wp-content/uploads/2019/03/Mining-royalties-research-report-final1.pdf (accessed 4 June 2019).

Dare, M., Schirmer, J. & Vanclay, F. (2014). Community engagement and social licence to operate. *Impact Assessment and Project Appraisal*, 32(3): 188–197.

Davenport, J. (2011). A brief history of the Jagersfontein diamond mine. *Mining Weekly* [online]. Available at www.miningweekly.com/article/a-brief-history-of-the-jagersfonteindiamond-mine-2011-11-11-1 (accessed 18 June 2017).

DRDLR (Department of Rural Development & Land Reform) (2016). District Rural Development Plan Sekhukhune District Municipality, Limpopo Province [online]. Available at www.ruraldevelopment.gov.za/services/geo-spatial-services-technology-and-rural-disaster/district-rural-development-plan-drdp/file/5884-sekhukhune-district-municipality-rural-development-plan (accessed 23 August 2018).

Edwards, P., Fleming, A., Lacey, J., Lester, L. et al. (2018). Trust, engagement, information and social licence – Insights from New Zealand. *Environmental Research Letters*, in press.

Fukuyama, F. (2011). *The origins of political order: From prehuman times to the French revolution.* New York: Farrar, Straus and Giroux.

Gamagara Development Trust (2013). GDF brings health services to the community of Gamagara with Re A Fola health promotion [online]. Available at www.gdf-trust.co.za/fola.php (accessed 22 July 2019).

Ga-Segonyana Local Municipality (2019). 2019–2020 Integrated development plan Ga-Segonyana Local Municipality [online]. Available at segonyana.gov.za/downloads/2019–20 FINAL IDP.pdf (accessed 22 July 2019).

Hathaway, O.A. (2003). Path dependence in the law: The course and pattern of legal change in a common law system. *John M. Olin Center for Studies in Law, Economics, and Public Policy Working Papers.* Paper 270 [online]. Available http://digitalcommons.law.yale.edu/lepp_papers/270 (accessed 14 June 2017).

Hoo, S.O. (2010). Anger over De Beers mine deal. *Diamond Field Advertiser*, 30 September, p. 13.

ICT (Itumeleng Community Trust) (n.d.). Itumeleng Community Trust, Jagersfontein [online]. Available at http://itumeleng-trust.org/ (accessed 18 June 2017).
ICT (Itumeleng Community Trust) (2015). Newsletter, February [online]. Available at http://itumeleng-trust.org/wp-content/uploads/2015/03/english-final.pdf (accessed 18 June 2017).
IGOPP (Institute for Governance for Private and Public Institutions) (2018). Board members are independent but are they legitimate and credible? [online] Available at https://igopp.org/en/board-members-are-independent-but-are-they-legitimate-and-credible/ (accessed 25 July 2019).
James, N. (2018). Perceptions mines are not doing enough for nearby communities fuelling protests. *Mining Weekly* [online]. Available at www.miningweekly.com/article/perceptions-mines-are-not-doing-enough-for-nearby-communities-fuelling-protests-2018-07-20 (accessed 29 March 2019).
Jaskoski, M. (2014). Environmental licensing and conflict in Peru's mining sector: A path-dependent analysis. *World Development*, 64(c): 873–883.
Joyce, S. & Thompson, I. (2000). Earning a social licence to operate: Social acceptability and resource development in Latin America. *Canadian Mining Metallurgical Bulletin*, 93(1037): 49–52.
Kingston, M. (2018). How to rebuild trust and integrity in South Africa [online]. Available at www.weforum.org/agenda/2018/06/how-to-rebuild-trust-and-integrity-in-south-africa/ (accessed 29 March 2019).
Kirsch, S. (2014). *Mining capitalism: The relationship between corporations and their critics*. Berkeley: University of California Press.
Koivurova, T., Buanes, A., Riabova, L., Didyk, V., Ejdemo, T., Poelzer, G., et al. (2015). "Social license to operate": A relevant term in Northern European mining? *Polar Geography*, 38(3): 194–227.
Ledwaba, L. (2018). Lives in limbo as despair seeps into mining ghost town. *Mail & Guardian* [online]. Available at https://mg.co.za/article/2018-06-15-00-lives-in-limbo-as-despair-seeps-into-mining-ghost-town (accessed 27 June 2018).
Maddison, S. (2009). *Black politics: Inside the complexity of Aboriginal political culture*. Crows Nest: Allen & Unwin.
Mahoney, J. (2000). Path dependence in historical sociology, *Theory & Society*, 29(4): 507–548.
Mandela Institute (2017). *Public regulation and corporate practices in the extractive industry: A South-South advocacy report on community engagement*. Johannesburg: University of Witwatersrand.
Marais, L., Matebesi, Z., Mthombeni, M., Botes, L. & Van Rooyen, D. (2008). Municipal unrest in the Free State (South Africa): A new form of social movement? *Politeia*, 27(2): 51–69.
Marais, L., McKenzie, F.H., Deacon, L. & Nel, E. (2018). The changing nature of mining towns: Reflections from Australia, Canada and South Africa. *Land Use Policy*, 76.
Marshall, J. (2015). Contesting big mining from Canada to Mozambique [online]. Available at www.tni.org/stateofpower2015 (accessed 14 June 2019).
Matebesi, S.Z. (2017). *Civil strife against local governance: Dynamics of community protests in South Africa*. Toronto: Barbara Budrich.
Matebesi S.Z. & Botes, L.J. (2017). Party identification and service delivery protests in the Eastern Cape and Northern Cape, South Africa. *African Sociological Review*, 21(2): 81–99.
Matebesi, S.Z. & Marais L. (2018). Social licensing and mining in South Africa: Reflections from community protests at a mining site. *Resources Policy*, 59(December): 371–378.

Meesters, M.E. & Behagel, J.H. (2017). The social licence to operate: Ambiguities and the neutralization of harm in Mongolia. *Resources Policy*, 53(September): 274–282.

MiningNews (2017). Celebrating transformation where it's needed most. *MiningNews* [online]. Available at https://miningnews.co.za/2017/06/26/celebrating-transformation-where-its-needed-most/ (accessed 17 March 2017).

Mining Review Africa (2015). Community protests prevent employees from working at Atlatsa Bokoni mine [online]. Available at www.miningreview.com/top-stories/community-protests-prevent-employees-from-working-at-atlatsa-bokoni-mine/ (accessed 17 March 2017).

MiningWatch Canada (2019). Hundreds take to the streets to call for justice for communities abused by Canadian mining [online]. Available at https://miningwatch.ca/news/2019/5/4/hundreds-take-streets-call-justice-communities-abused-canadian-mining (accessed 14 June 2019).

Mining Weekly (2010). De Beers sells Jagersfontein mine to BEE consortium. *Mining Weekly* [online]. Available at www.miningweekly.com/article/de-beers-sells-jagersfontein-mineto-bee-consortium-2010-09-28/rep_id:3650 (accessed 18 June 2017).

Moffat, K., Lacey, J., Zhang, A. & Leipold, S. (2016). The social licence to operate: A critical review. *Forestry*, 89: 477–488.

Moffat, K. & Zhang, A. (2014). The paths to social licence to operate: An integrative model explaining community acceptance of mining. *Resources Policy*, 39(March): 61–70.

Motse, O. (2019). Jagersfontein mine, FS municipality to square off in High Court today. *OFM News* [online]. Available at www.ofm.co.za/article/centralsa/271118/jagersfontein-mine-fs-municipality-to-square-off-in-high-court-today (accessed 17 May 2019).

NCNN (Northern Cape News Network) (2018) [online]. Available at https://ncnn.live/jtg-developmental-trust-marks-community-development/ (accessed 17 May 2019).

Office of the Premier-NC (Northern Cape) (2012). Protests in John Taolo Gaetsewe District [online]. Available at www.northern-cape.gov.za/index.php/news-room/news-and-speeches/152-media-room/office-of-the-premier/news-and-speeches/779-acting-premier-speech-protests-in-john-taolo-gaetsewe-district (accessed 14 May 2018).

Owen, J. & Kemp, D. (2013). Social licence and mining: A critical perspective. *Resources Policy*, 38(1): 29–35.

Owen, J. & Kemp, D. (2014). Mining and community relations: Mapping the internal dimensions of practice. *Extractive Industries and Society*, 1(1): 12–19.

Owen, J. & Kemp, D. (2017). *Extractive relations: Countervailing power and the global mining industry*. New York: Routledge.

Petrov, A. (2016). Exploring the Arctic's "other economies": Knowledge, creativity and the new frontier. *The Polar Journal*, 6(1): 51–68.

Pickering, J. & Nyapisi, T. (2017). A community left in the dark: The case of Mapela. Struggles for transparency and accountability in South Africa's Platinum Belt. In: *Good company* (2nd edn, pp. 27–35). Conversations around transparency and accountability in South Africa's extractive sector by Open Society Foundation for South Africa. OSF: Cape Town.

Prno, J. & Slocombe, D.S. (2012). Exploring the origins of "social license to operate" in the mining sector: Perspectives from governance and sustainability theories. *Resources Policy*, 37(3): 346–357.

Riofrancos, T. (2019). What comes after extractivism? *Dissent Magazine* [online]. Available at www.dissentmagazine.org/article/what-comes-after-extractivism (accessed 29 June 2019).

Seccombe, A. (2010). De Beers sells assets of defunct operation. *Business Day*, p. 13 [online]. Available at https://allafrica.com/stories/201009290636.html (accessed 23 July 2019).

Smith, C., 2010. Aansoeke sluit vandag om diamante te delf by Jagersonfontein-myn. *Volksblad* 2.

Solomon Star (2018). Trust restoring the dignity of the people [online]. Available at https://solomonstar.live/trust-restoring-dignity-of-the-people/ (accessed 12 November 2018).

SowetanLive (2011). DNA results deepen chieftaincy row [online]. Available at www.sowetanlive.co.za/news/2011-09-13-dna-results-deepen-chieftancy-row/ (accessed 17 June 2016).

StatsSA (Statistics South Africa). (2011). *Community survey, 2007.* Pretoria: Statistics South Africa.

Taylor, D. & Mahlangu, S. (2017). *Earning the social licence to operate: A case study about culture.* Proceedings of the European Conference on Management, Leadership & Governance [online]. Available at https://eds.a.ebscohost.com/eds/pdfviewer/pdfviewer?vid=4&sid=2dc9-df10-f4c2-45d7-af7f-f002499a1d70%40sdc-v-sessmgr01b (accessed 12 June 2018).

Thomson, I. & Boutilier, R. (2011). The social licence to operate. In: *SME mining engineering handbook* (pp. 1779–17796). Englewood, CO: Society for Mining, Metallurgy and Exploration.

Tilly, C. (2004). *Trust and rule.* New York: Columbia University Press.

Uslaner, E.M. (2018). The study of trust. In: E.M. Uslaner (ed.). *The Oxford handbook of social and political trust* (pp. 3–14). New York: Oxford University Press.

Van Assche, K., Beunen, R. & Duineveld, M. (2014). *Evolutionary governance theory: An introduction.* Heidelberg: Springer.

Vanclay, F. & Hanna, P. (2019). Conceptualizing company response to community protest: Principles to achieve a social license to operate. *Land*, 8(6): 101.

Veiga, M., Scoble, M. & McAllister, M.L. (2009). Mining with communities. *Natural Resources Forum.*

Walsch, B., Van der Plank, S. & Behrens, P. (2017). The effect of community consultation on perceptions of a proposed mine: A case study from southeast Australia. *Resources Policy*, 51(March): 163–171.

6 Conclusion

Social licensing and prospects for community development

Introduction

One of the most important questions facing resource-rich countries globally today is how to optimise natural-resource revenues for the long-term sustainable development of mining-affected communities. It is thus not surprising that the failure by national development plans to address the governance of extraction by mining-affected communities ranks among the primary manifestations of conflict related to mining activities. This conflict is further exacerbated by the twofold role of mining in society: mining not only sustains population wellbeing and economic growth, but also generates social and environmental impact, which may jeopardise the acceptance of mining projects. The twofold role of mining is aptly espoused by scholars advancing the resource curse debate that naturally resource-rich countries tend to grow more slowly than resource-poor countries.

Companies in the resource sector are involved in many initiatives in an attempt to deal with the perceived public legitimacy and acceptance of mining activities. The concept of corporate social responsibility (CSR) is entrenched in the mining industry, with many mining companies having adopted the social licence to operate (SLO) model as complementary (Thomson & Boutilier, 2011; Wilburn & Wilburn, 2011; Zhang et al., 2015). As a result, the SLO has become ubiquitous in the extractive sector. This ubiquity of the SLO led, however, to the oscillation between those who hail an SLO as an important mechanism to build constructive relationships with mining-affected communities and those who believe it is impossible to implement an SLO in practice. Notwithstanding this debate, the level of acceptance or approval continually granted to an organisation's operations or project by the local community and other stakeholders presents a profound threat to mining company–community relations. Of central importance, in this regard, is the role played by trust according to the four levels of an SLO: withdrawal, acceptance, approval, and psychological identification (Thomson & Boutilier, 2011).

There is a wealth of research which has generated several analytical perspectives on SLO (Brereton et al., 2011; O'Faircheallaigh, 2012; Bice, 2014; Moffat & Zhang, 2014; Keenan et al., 2016; Loutit et al., 2016; Owen, 2016; Meesters & Behagel, 2017). However, few studies have been conducted on community development trusts (CDTs) in South Africa. Drawing on four case studies of CDTs – an informal community-based resource management structure – from four provinces of South Africa, the study sought to examine how an SLO is obtained and maintained through the lenses of path dependency within evolutionary governance theory (EGT) (Beunen et al., 2016). These CDTs serve as implementing units for mining grant and CSR projects.

In particular, the cases yield mixed results in respect of how path dependency is created, maintained, or disrupted. In this regard, a major theme of the book has also been how has Royal Bafokeng Nation (RBN) managed to become a model of success for community-based management of mining revenues. This case demonstrates how successive generations of leadership decisions and actions of the RBN – underpinned by reconciling traditional and modern governance institutions – created and maintained a profound path-dependent system. Furthermore, in stark contrast with many other CDTs under traditional governance – in fact, CDTs across the board – it highlights the key opportunities of effective linkages between traditional and modern institutions of governance.

This chapter is divided into six sections. This introductory section provided a backdrop to the chapter. The next section discusses customarily law and efforts to enforce an SLO, thereby illustrating the relationship between the policy level and the decision-making levels in path dependency. "The unique case of the Royal Bafokeng" section looks at the RNB case, followed by a general examining of CDTs in the next. The last section concludes.

Civil society and local ownership – a toxic residue

Considerable attention has been devoted to the role of civil society in advancing the interests of mining-affected communities. This section tries to situate the debate within the context of the nature of South African civil society. South Africa has a vibrant civil society that played a pivotal role in the domestic struggle against apartheid. Despite being poorly coordinated and scantly funded, it still wields considerable influence (Oosthuizen, 2019).

Obtaining an SLO about extraction projects is more relevant than ever. Globally, public awareness concerning the social and environmental impact of mining projects is constantly increasing, resulting in heightened conflict between mining companies and mining-affected communities. While the genesis of such conflicts is complex, there is no question about the significance of the quest of communities around the world to assert their rights and role in decision-making processes about mining activities where they live. In this regard, communities like the Munduruku people in the Brazilian Amazon, Lamu in Kenya, Maya Qeqchi in Guatemala, the Pinuyamayan of Kasavakan in Taiwan, and the AbaThembu in Xolobeni in South Africa have successfully managed to safeguard their environments.

Within the current context of popular protests in South Africa, the increasing advocacy role of civil society organisations (CSOs) in advancing the rights of precarious mining-affected communities highlights the complicated relationship between the government and civil society. As one of the key stakeholders, according to the SLO conceptual framework outlined in Chapter 1, CSOs serve as alternative avenues to weak institutions and provide marginalised communities with a platform on which they can organise themselves and the opportunity to influence the development trajectory of their communities (Manyuchi, 2012).

The analyses in this book, while supporting the critique in the literature that communities are not necessarily unified structures (Matebesi & Marais, 2018), argue that this concern is overstated. This raises theoretical concerns related to issues of legitimacy, democracy, and internal conflict, considering that even highly heterogeneous societies may have different competing interest groups. Thus, I argue that questions around the unitary state of communities ignore the plurality of actors within society and raise concerns related to issues of legitimacy among CSOs. The findings have shown that this has encouraged some mining companies to question the standing of grassroots community-based activist groups instead of dealing with their fundamental grievance, as in the cases of Burgersfort and Jagersfontein.

Popular community mobilisation by mining-affected communities primarily centred on land rights and engagement has gained traction in South Africa. These protests are not confined to urban settlements and have spread to places that had not experienced this phenomenon before. However, the success of this protest is varied. Evidence from the present book shows that it is very difficult, if not impossible, for grassroots activist groups to engage in successful mobilisation against the government and mining companies without the support of national interest groups.

A major challenge for these grassroots organisations is that they have to operate within a macro structural framework and negotiate with the government and mining companies. As in the case of Greenen and Verweijen (2017), the South African political and socio-economic context, characterised by intense conflicts and patronage-based politics, limits the mobilisation potential of grassroots organisations. It is also evident that mining–company practices, in particular, the co-optation of intermediaries and protestors, and creation of a repressive climate are central to social mobilisation in mining concessions.

Scholars agree on the central role of national organisations as a response to the deficit in decision-making and participation processes (Arce, 2014). As we have noticed in the study, Mining Communities United in Action (MACUA), together with a national network of other CSOs, secured significant benefits for mining-affected communities. As in the case of countries such as Niger Delta, the Philippines, and Russia, a recent report by non-profit environmental justice service and developmental organisations titled "We Know Our Lives Are in Danger" (Human Rights Watch, 2019) highlights widespread incidences of harassment, intimidation, and the killing of activists who campaign against mining projects.

Despite these challenges, CSOs in South Africa have enabled marginalised communities like Lesetlheng and Xolobeni to win high-profile cases that will fundamentally change the power balance between mining companies and communities. This reinforcement of the role of communities in mining decision-making processes is especially vital at a time when the government seems to uphold its hegemonic statutory rights to natural resources and in promoting the business-centric goal of the extractive industry.

In the South African case, questions have to be asked about the current prospects for the continued involvement of CSOs in supporting mining-affected communities. In this regard, Oosthuizen (2019) stated:

> It has become clear that South African politicians will not, of their own volition, serve the national interest. It's also apparent that SA's business community will not demonstrate an enduring commitment to the long-term future of the country in the face of dysfunctional politics. What arises from these two immutable observations is that the country's future lies in the hands of civil society.

And Langman and Benski (2019: 318) added:

> Accordingly, in many cases, right-wing, conservative leaders and constituencies are not only resentful, perhaps fearful, toward progressives and progressive agendas but the very emergence of such movements often dispose of virulent reactionary social movements that would halt, if not reverse, social change and end such movements.

More concerning is that there are increasing cases of strategic litigation against public participation (SLAPP), legal bullying tactics by corporations and governments to silence civic activists by threatening them with hefty damage claims and protracted litigation (Carnie, 2019). However, civil society in South Africa is fighting back against abusive corporate tactics through a new joint advocacy campaign known as *Asina Loyiko* (We Have No Fear) – United Against Corporate Bullying (Fourie, 2019).

The social licence and customary law in South Africa

Globally, there is a significant diversity of the existing agreement or models of the distribution of benefits ensure the sustainable development of local communities, which include paternalism, CSR, and partnerships (Burtseva & Bysyina, 2019). In South Africa, significant regulatory progress has been made in respect of the legal licence to operate in the form of social and labour plans, which dictate that mines have to consider the socio-economic and environmental wellbeing of their employees and the local community where they operate (Centre for Applied Legal Studies, 2017).

In reality, many mining companies in South Africa do not fulfil the obligations set out in their social and labour plans (Centre for Applied Legal Studies, 2017). If mining companies fail to comply with legal requirements, the question remains how they will fulfil their social responsibilities to society. Consequently, systemic barriers prevent communities from participating in extractive governance decision-making processes that affect them meaningfully.

A major factor that encourages mining companies to operate without seeking the consent of mining-affected communities is the mining regulations in South Africa that seem to override the legal protection for informal land rights. This is the case with the continued practice whereby the government grants licences to mining companies on land governed by customary law without seeking the consent of the communities living on the land.

With the current weak legislative oversight from the ministry of mineral resources in South Africa, there seems to be no indication there will be any time soon a change in commitment to remove these systemic barriers. For example, almost a third of South Africans still live on the communal land that made up the former apartheid homelands. A protest to Pretoria in June 2019 led by the Alliance for Rural Democracy (representing rural communities from seven provinces) demanded President Cyril Ramaphosa not sign into law two controversial bills: the Traditional Courts Bill (also known as the Bantustans Bill) and the Traditional Khoisan Leadership Bill (Mabasa, 2019).

Clause 24 of the Traditional Khoisan Leadership Bill stipulates that traditional councils can enter into agreements with third parties without obtaining the consent of the community. While it does make provision for "consultation," there is no requirement for consent from the community (Mabasa, 2019). In the context of this study, this means that traditional leaders can conclude mining deals without the consent of their subjects. I argue that these explicit discriminatory laws effectively empower traditional leaders while disempowering people living there. This raises another possible concern about why politicians have ratified the traditional bills.

While traditional healers may be socially accountable and serve as development brokers (Baldwin and Raffler, 2017), this is largely not cased in South Africa. A plausible explanation perhaps is that powerful traditional leaders act as intermediaries or valuable economic and political partners (Baldwin, 2016) and "have incentives to support incumbent political parties who can guarantee their survival and provide them with rents" (De Kadt & Larreguy, 2018). Given these hurdles, the credibility and legitimacy of traditional leaders to conclude mining deals on behalf of communities will fuel ongoing protests in traditional communities where extractive operations take place.

The unique case of the Royal Bafokeng

This section synthesises the empirical findings to answer the study's research questions relating to the governance and impact of CDTs on local communities. It also attempts to outline the factors that enabled the RBN Development Trust to be successful, while other CDTs are failing. As the case studies illustrate, mining companies use CDTs, often named after a town or ethnic group. Evidence of international studies has shown that institutional frameworks of extraction have different effects on, for example, societies and economies (Costanza, 2016;

Conde & Le Billon, 2017; Dagvadorj et al., 2018; Burtseva & Bysyina, 2019). There is also a consensus in the literature that quality institutions and accountability mechanisms can generate positive results from natural-resource extraction (Ahmadov & Guliyev, 2016).

Another important dimension of the SLO conceptual framework is the quality of consultation. As we saw in Chapter 5, discussions with community leaders in the study sites revealed that most of them did know how CDTs were established. However, they emphasised that many companies accept the need to apply for the SLO through community engagement, but often devise means to circumvent their responsibility towards communities. A major concern of those interviewed was that CDTs are used as vehicles to enrich certain individuals, while the community is provided information via websites or annual reports.

The findings reveal that the board of trustees of these trusts are in most cases the local political elite. In the case of Kuruman, the trust did not have any community meeting for ten years. Interestingly, it is also evident how core leaders of the mobilisation campaigns of mining-affected communities – in line with the African adage "An empty stomach has no ears" – secretly tried to conclude personal deals when engaging with representatives of mining companies. This situation demonstrates that when competing actors – mining companies, government, and mining-affected communities – seek to legitimise the rational use and governance of resources, there are bound to be conflicting priorities. In its current manifestation, CDTs and, by default, the social licence come across more as a product of these conflicting priorities than community agency. Moreover, finding a balance seems to be elusive for CDTs. This impacts on the legitimacy and credibility of mining operations.

As we saw in Chapter 4 (and many other academic and media reports, for example; Mbenga & Manson, 2010; Totem Media, 2011), the RBN traditional authority exerted its land rights through legal processes. Using the options available within customary and modern institutional governance frameworks, which are linked by the RBN Development Trust, the Bafokeng traditional authority worked towards protecting the interests of the Bafokeng over many decades. At the end of 2017, the Royal Bafokeng Holdings' portfolio grew to R32 billion (RBH, 2017). RBN has been able to use platinum revenue for community development through the provision of several services, including education, health, and infrastructure.

Interestingly, the traditional governance system largely remains patriarchal, while the modern sphere consists of the administration sphere, and employment is based on the principles of equality and fairness. Within this model, RBN concludes all the business deals of it on behalf of the nation. However, community engagement has been a key strategy of accountability. For example, all entities and funds are annual, while financial statements and project reports are presented at community meetings.

Within the traditional governance system, traditional leaders are responsible for consulting with local communities through an elaborate customary consultative process. This confirms several studies on the social responsibility of mining, which identify the relationships with local communities and stakeholders, including the dissemination of quality social reports, as essential to mining companies in acquiring local legitimacy from mining-affected communities. These measures appeared successful in finding a balance between local socio-political and cultural power structures, as well as building and maintaining the trust of RBN.

Despite these achievements, there have been two fundamental questions about the success of RBN. First, the questions relate to issues of mutual benefits, equitable distribution of communal resources, and evidence of service delivery (Cook, 2005; 2011, 2013). Several community members echoed this concern, accusing the tribal authority of nepotism, corruption, greed, and poor management of extractive resources. A more critical view from local perspectives is the growing belief that the pervasive sense of entitlement enshrined in the customary governance practices has no place in a democratic South Africa.

The second question is whether the success of RBN can be ascribed to property land rights (which give it access to a unique abundance of natural resources) or the resource governance model is adopted. Notably, communal land rights gave RBN ownership rights to minerals, but ownership in itself is insufficient in addressing the multifaceted challenges of local resource governance. Another useful point is that successful natural-resource governance depends, to a large extent, on the design of governance arrangements (Owen & Kemp, 2013, 2014, 2017; Keenan et al., 2016), which requires normative guidance. Principles that can be used to direct the design of governance institutions like those proposed by Lockwood et al. (2010) include legitimacy, transparency, accountability, inclusivity, and fairness. I now look at the main findings in respect of other traditional and modern CDTs.

Challenges facing CDTs in community-based natural-resource governance

An analysis of the history of CDTs in South Africa demonstrates that these structures arose out of the desire of mining companies in South Africa to expand their philanthropic footprint. As has been shown in this book, there have been widespread protests by mining-affected communities over the benefits of natural resources. There has not been much difference between the CDT led by customary leaders in Atok, Limpopo, and the two others in Kuruman (Northern Cape) and Jagersfontein (Free State).

While the CDT process reflects elements of social acceptance, the number of complaints against them, particularly about a lack of transparency, has raised several concerns about the legitimacy of these structures. For example, communities in the case studies complained that they are consulted on rare occasions (in one case, only after 15 years) when there is no detail about how proceeds from mines will be used. Significant progress seems to have been made in other cases by having a legally constituted board of trustees, consisting of representatives of the mine, the community, and other stakeholders. However, the community members view these arrangements with trepidation. They point to the tokenistic inclusion of community representatives, who are often susceptible to secret arrangements with the trusts.

Struggles over customary leadership – rooted in a scramble for access to mining contracts – have been a significant barrier to socio-economic development in some mining-affected areas. In the case of Atok, where more than ten mines operate, several villages have no direct access to clean and safe drinkable water, despite the trust having received R2.4 million in contributions from the local mine. The tribal authority has failed to provide financial statements and provide the community with the trust deed. As a result, the Atok area has experienced high incidences of protests against mining operations. These entrenched trends in CDTs, under both customary and modern leadership, could challenge their ability to emulate the success of RBN. A more pertinent challenge, however, will depend on how all actors – industry, government, and local stakeholders – subject CDTs to modern corporate management practices.

However, while mining-affected communities have the right to demand to be involved in the decision-making processes of natural-resource governance, CDTs should not be seen as a means to righting past developmental injustices. However, equally so, CDTs should also not address some issues in communities and assume that an SLO has been obtained. Meanwhile, the RNB case – while far from being perfect – demonstrates how path dependency was created and maintained, leading to the success of the Tswana nation. CDTs could take a cue from the RNB case and invest in community engagement to break path dependency.

Concluding observation

Finding effective solutions for natural-resource governance is, in any context, a complex challenge compounded by the different intrinsic needs of the government, industry, and mining-affected communities. It is generally accepted that achieving social acceptance and high levels of legitimacy among all stakeholders that are directly affected by development projects is significant for the proper and effective implementation of those projects. The experience from several cases around the world, including in this book, has demonstrated that stakeholders are more likely to accept mining operations in their region if mining companies are perceived as being fair in their decision-making processes. By an extension, such a normative expectation from mining-affected communities and civil society at large is fundamentally concerned with community-level impact.

Accordingly, SLO processes are by no means an attempt to impede the ability of the government to exercise its statutory responsibility in respect of natural-resource management. I acknowledge the responsibility of governments to create and enforce mining-related policies. Similarly, it will be unfair to expect the extractive industry to solve long-standing chronic problems faced by many mining-affected communities across the world. In the context of South Africa, leading mining companies like Anglo American SA, Amplats, Coal SA, De Beers Consolidated Mines, and Kumba Iron Ore should be commended for running what has been described as possibly the world's most sophisticated and complex social investment operation.

Notwithstanding this, I believe, we can also assume that it is highly likely that mining-affected communities will counter any deliberate attempt to exclude them from localised natural-resource governance processes. Furthermore, the exclusion of representatives of mining-affected communities from discussions of Mining Charters, which related to employment equity, mine community and rural development, and housing and living conditions, was bound to be challenged by CSO mobilisation in the country. Such mobilisation by CSOs should be seen as a legitimate means to curtail the government and industry's capacity to exercise hegemonic influence over the scope and shape of social investment programmes in the name of, but without, mining-affected communities.

Despite what theoretical, policy, and industry debates report about the benefits of mining, in reality, the persistent and severe negative consequences of mining on the environment, health, and the social fabric of community's health relate a different story. The benefits of mining revenues have proven to be comprehensive and sustainable only when localised resource governance models follow transparent and accountable institutional governance practices. Given these outcomes and the case studies presented, concerted measures should be taken for a prudent localised natural-resource governance strategy in South Africa. Procedural fairness in the management of CDTs is central to the disrupting path dependency and achieving broad and localised societal acceptance of mining operations. In this manner, South Africa and the many other countries around the world that follow localised resource management processes will achieve their transformative development objectives.

References

Ahmadov, A.K. & Guliyev, F. (2016). Tackling the resource curse: The role of democracy in achieving sustainable development in resource-rich countries. *International IDEA Discussion Paper.*

Arce, M. (2014). *Resource extraction and protest in Peru.* Pittsburgh, PA: University of Pittsburgh Press.

Baldwin, K. (2016). *The paradox of traditional chiefs in democratic Africa.* New York: Cambridge University Press.

Baldwin, K. & Raffler, P. (2017). Traditional leaders service delivery and electoral accountability. In: J. Rodden & E. Wibbels (eds). *Decentralization and development in practice: Assessing the evidence.* New York: Cambridge University Press.

Beunen, R., Van Assche, K. & Duineveld, M. (2016). *Evolutionary governance theory.* New York: Springer.

Bice, S. (2014). What gives you a social licence? An exploration of the social licence to operate in the Australian mining industry. *Resources,* 3(1): 62–80.

Brereton, D., Owen, J. & Kim, J. (2011). *Good practice note: Community development agreements.* EI Source Book. Dundee: Centre for Energy, Petroleum and Mineral Law and Policy (CEPMLP); University of Dundee.

Burtseva, E. & Bysyina, A. (2019). Damage compensation for indigenous peoples in the conditions of industrial development of territories on the example of the Arctic Zone of the Sakha Republic. *Resources,* 8(5): 1–14.

Carnie, T. (2019). Mining company threatens to 'Slapp' green activists over Facebook posts [online]. Available at www.dailymaverick.co.za/article/2019-04-25-mining-company-threatens-to-slapp-green-activists-over-facebook-posts/ (accessed on 30 July 2019).

Centre for Applied Legal Studies (2017). *The social and labour plan series phase 2: Implementation operation analysis report.* Johannesburg: University of Witwatersrand.

Conde, M. & Le Billon, P. (2017). Why do some communities resist mining projects while others do not? *Extractive Industries and Society,* 4(3): 681–697.

Cook, S.E. (2005). Chiefs, kings, corporatization and democracy: A South African case study. *Brown Journal of World Affairs,* 12(1): 125–137.

Cook, S.E. (2011). The business of being Bafokeng: The corporatization of tribal authority in South Africa. *Current Anthropology,* 52(S3): S151–S159.

Cook, S.E. (2013). Community management of mineral resources: The case of the Royal Bafokeng Nation. *Journal of the Southern African Institute of Mining and Metallurgy,* 113(1): 61–66.

Costanza, J.N. (2016). Mining conflict and the politics of obtaining a social license: Insight from Guatemala. *World Development,* 79(C): 97–113.

Dagvadorj, L., Byamba, B. & Ishikawa, M. (2018). Effect of local community's environmental perception on trust in a mining company: A case study in Mongolia. *Sustainability,* 10: 614.

De Kadt, D. & Larreguy, H.A. (2018). Agents of the regime? Traditional leaders and electoral politics in South Africa. *Journal of Politics,* 80(2): 382–399.

Fourie, M. (2019). Civil society unites against corporate censorship and bullying [online]. Available at www.dailymaverick.co.za/opinionista/2019-05-29-civil-society-unites-against-corporate-censorship-and-bullying/ (accessed 30 July 2019).

Greenen, S. & Verweijen, J. (2017). Explaining fragmented and fluid mobilization in gold mining concessions in the eastern Democratic Republic of the Congo. *The Extractive Industries and Society,* 4(4): 758–765.

Human Rights Watch (2019). "We know our lives are in danger": Environment of fear in South Africa's mining-affected communities [online]. Available at www.hrw.org/report/2019/04/16/we-know-our-lives-are-danger/environment-fear-south-africas-mining-affected (accessed 30 July 2019).

Keenan, J.C., Kemp, D.L. & Ramsay, R.B. (2016). Company-community agreements, gender and development. *Journal of Business Ethics,* 135(4): 607–615.

Langman, L. & Benski, T. (2019). Global justice movements: Past, present, and future. In: B. Berberoglu (ed.). *The Palgrave handbook of social movements, revolution, and social transformation* (pp. 301–324). Cham: Palgrave McMillan.

Lockwood, M., Davidson, J., Curtis, A., Stratford, E. & Griffith, R. (2010). Governance principles for natural resource management. *Society & Natural Resources,* 23(10): 986–1001.

Loutit, L., Mandelbaum, J. & Szoke-Burke, S. (2016). Emerging practices in community development agreements. *Journal of Sustainable Development. Law & Policy,* 7(1): 64–96.

Mabasa, N. (2019). People take the fight against traditional leaders to president Ramaphosa. *Daily Maverick* [online]. Available at www.dailymaverick.co.za/article/2019-06-05-people-take-the-fight-against-traditional-leaders-to-president-ramaphosa/.

Manyuchi, R. (2012). *The role of civil society organisations/non-governmental organisations (CSOs/NGOs) in building human capability through knowledge construction: The case of Africa community publishing development trust (Zimbabwe).* Doctoral thesis. Stellenbosch: University of Stellenbosch.

Matebesi, S.Z. & Marais L. (2018). Social licensing and mining in South Africa: Reflections from community protests at a mining site. *Resources Policy*, 59(Dec 2018): 371–378.

Mbenga, B. & Manson, A. (2010). *"People of the dew": A history of the Bafokeng of Phokeng-Rustenburg Region, South Africa, from early times to 2000*. Auckland Park: Jacana.

Meesters, M.E. & Behagel, J.H. (2017). The social licence to operate: Ambiguities and the neutralization of harm in Mongolia. *Resources Policy*, 53(September): 274–282.

Moffat, K. & Zhang, A. (2014). The paths to social licence to operate: An integrative model explaining community acceptance of mining. *Resources Policy*, 39(March): 61–70.

O'Faircheallaigh, C. (2012). Community development agreements in the mining industry: An emerging global phenomenon. *Community Development*, 44(2): 222–238.

Oosthuizen, M. (2019). SA civil society: The road less travelled. *Daily Maverick* [online]. Available at www.dailymaverick.co.za/opinionista/2019-06-09-sa-civil-society-the-road-less-travelled/ (accessed 11 June 2019).

Owen, J.R. (2016). Social license and the fear of Mineras Interruptus. *Geoforum*, 77: 102–105.

Owen, J. & Kemp, D. (2013). Social licence and mining: A critical perspective. *Resources Policy*, 38(1): 29–35.

Owen, J. & Kemp, D. (2014). Mining and community relations: Mapping the internal dimensions of practice. *Extractive Industries and Society*, 1(1): 12–19.

Owen, J. & Kemp, D. (2017). *Extractive relations: Countervailing power and the global mining industry*. New York: Routledge.

Royal Bafokeng Holdings. (2017). RBH integrated review 2017 [online]. Available at www.bafokengholdings.com/images/pdf/annual-review-2017.pdf.

Thomson, I. & Boutilier, R. (2011). The social licence to operate. In: P. Darling (ed.). *SME mining engineering handbook* (pp. 1779–1796). Englewood, CO: Society for Mining, Metallurgy and Exploration.

Totem Media (2011). *Mining the future the Bafokeng story*. Auckland Park: Jacana Media.

Wilburn, K.M. & Wilburn, R. (2011). Achieving a social license to operate using stakeholder theory. *Journal of International Business Ethics*, 4(2): 3–16.

Zhang, A., Moffat, K., Lacey, J., Wang, J., González, R., Uribe, K., Cui, L. & Dai. Y. (2015). Understanding the social licence to operate of mining at the national scale: A comparative study of Australia, China and Chile. *Journal of Cleaner Production*, 108(Part A): 1063–1072.

Index

abuses by mining companies 110
Abuya, W.O. 67
ACC *see* Amadiba Crisis Committee (ACC)
Adali, S. 28
Africa: customary tenure in 61; foreign investments 62; mining regulation in 60–2
Africa Mining Vision (AMV) 62
African National Congress (ANC) 4, 56, 57
Alliance for Rural Democracy 143
Alternative Mining Indabas 5, 62
Amadiba Crisis Committee (ACC) 72
amoral familism 26–7
Ampofo-Anti, A.O. 94
AMV *see* Africa Mining Vision (AMV)
Andersson, K. 37
Anglo American Chairman's Fund 65
Anglo American (2019) report 65
Annooraq Community Participatory Trust 114–19
anti-extractive movements 66
Article 20 of the Constitution of 1992 61
Asina Loyiko 142
Atok case, social mobilisation against CDTs 17, 112, 146; Annooraq Trust 114; BBNCEC's role 115; Bokoni mine 114; community engagement 116–18; community protests 115; DMR meetings 117; future pathway 118–19; local impact and benefits 118
Atok Community Development Association 117
Atok Community Engagement Forum 115
Atomic Energy Act 90 of 1967 **54**, 55
Australia 1, 38; compensation for land access and acquisition 7; issue of aboriginal rights 53; mining regulatory framework 62; mining royalties in 5

Bafokeng Land Buyers' Association (BLBA) 93–4
Banfield, E.C. 26, 27
Bantustans Bill *see* Traditional Courts Bill
Baroka Ba Nkwana Community Engagement Committee (BBNCEC) 114–18
Baroka Ba Nkwana Community Trust (BBNCT) 112
Baroka tribe 112
Base Minerals Development Act 39 of 1942 **54**
BBBEE *see* Broad-Based Black Economic Empowerment (BBBEE)
BBNCEC *see* Baroka Ba Nkwana Community Engagement Committee (BBNCEC)
BBNCT *see* Baroka Ba Nkwana Community Trust (BBNCT)
Bebbington, A. 111, 125, 132
BEE Commission Report (2001) 57
BEE laws 61
benefit-sharing agreements 39, 83
Benski, T. 142
Bentley blockade 35
Benton, N. 33
Bice, S. 11
Black Land Act 27 of 1913 56
black mineworkers 3
BLBA *see* Bafokeng Land Buyers' Association (BLBA)
Bodo community 66–7
Bokoni mine: closure of 114; protests against 115; reopening of 117–19
Bolivia 132; Tacana community in 67
Booi, Z. 75
Botes, L.J. 74–5
Boutilier, R.G. 9, 10

Index

Brakfontein Klipfontein 116
"Breaking New Ground: Mining, Minerals and Sustainable Development" 9
Broad-Based Black Economic Empowerment (BBBEE) 57
Broad-Based Socio-Economic Empowerment Charter for the South African Mining and Minerals Industry (2018) 2
"brown envelope" phenomenon 59
Bunker, S. 6, 111, 132
Bury, J. 111, 125, 132

Canada 1; compensation for land access and acquisition 7; local-level agreements in 64; mining industry in 63
Carte, D. 98
Cawood, F.T. 56, 86
CDAs *see* community development agreements (CDAs)
CDTs *see* community development trusts (CDTs)
Chamber of Mines 69, 70
Changes to the South African Mineral and Petroleum Resources Royalty Act of 2008 ("Royalty Act") **58**
Chhotray, V. 12
Chimhowu, A. 61
civil society mobilisation against mining 66–7, 76, 147; core issues 67; CSO response to exclusion to mining agreements 69; CSR-driven initiatives for 67; indigenous Shuar activists 67; mining charters and court challenges 69–71; in South Africa 67–8; Xolobeni and Lesetlheng communities 71–5
civil society organisations (CSOs) 3, 41, 53, 140–2; advocacy role of 140; exclusion of communities from mining laws 69; mining charters and court challenges 69–71; in South Africa 141
Clapham, D. 5
Cohen, M.A. 28, 29
communal land rights 145
Communal Land Rights Act (No. 11 of 2004) **87**, 88
community acceptance of mining 1; community-mine relations 31–3; impact on social infrastructure 30–1; integrative model of 10–11, 16, 26; procedural fairness and 33–6
community-based natural-resource governance 83, 145–6

community development agreements (CDAs) 7, 14
community development trusts (CDTs) 1, 14, 33, 43, 110, 133, 139–40, 144; analysis of 8; case studies of 15; challenges 65–6, 145–6; contribution to quality of life 130, **131**; history of 145; as local-level agreements 6–9; procedural fairness and trust in 16, **130**; theories of elite capture 36–41; *see also* social mobilisation against CDTs
community engagement 33, 38, 144; in Atok case 116–18; in consultation process 92–5; in Jagersfontein case 121; in Kuruman case 126–7; legal and regulatory frameworks for 59–63
community–mine relations: harmful contact 32; Indonesian study on 39; positive contact 31–3; in South Africa 3–6; trust and distrust 10–11
community protests 53, 68, 74, 103; against Atok/Bokoni mine 115, 146; in Jagersfontein 120; in Kuruman 113; against Mogalakwena platinum mine 35
community trusts *see* community development trusts (CDTs)
compensation 56, 57, 72; for land access and acquisition 7
competency-based trust 28
Constitution of the Republic of South Africa 57
Convention 169 *see* International Labour Organization (ILO) Convention 169
Cook, S.E. 84, 86, 88, 89, 93
corporate social responsibility (CSR) 8, 9, 26, 139
Corruption Watch report 83–4, 111
credibility 10, 120, 128
CSR *see* corporate social responsibility (CSR)
customary law: in African countries 61; in South Africa 39, 142–3
customary leadership 146
customary tenure 61

Daily Maverick (2016) 34
Dale, M.O. 55
Davenport, J. 6
De Beers 113
decision-making processes 111; public involvement in 3–4, 33, 53; of RNB 104
Deloitte (2018) 60
Demmers, J. 37
democratic accountability 52

Index

Department of COGTA 85, 86
Department of Energy 53
Department of Mineral Resources (DMR) 34, 53, 58, 69, 72, 132; in Atok case 114–19
Department of Trade and Industry 65
Development Land and Trust Act 18 of 1936 56
Dienhart, J.J. 29
distrust of mining companies 10
DMR *see* Department of Mineral Resources (DMR)
Draft Reviewed Mining Charter 70
Duffy, S. 26, 42
Dumela Phokeng meetings 93

Eastern Europe, shale gas industry in 63
EGT *see* evolutionary governance theory (EGT)
EITI standard *see* Extractive Industries Transparency Initiative (EITI) standard
elite capture and SLOs 17, 36–7; definition of 37; influence elites 37; within local participatory processes 40–1; in mining industry 37–8; traditional elites 38–40
environmental impact assessment (EIA) 67
Etter-Phoya, R. 7
evolutionary governance theory (EGT) 3, 87, 92, 111, 140; and path dependency 12–13, 42
extractive governance approach 52
Extractive Industries Transparency Initiative (EITI) standard 64
Extractive Relations: Countervailing power and the global mining industry (Owen and Kemp) 11, 29

Fanon, Frantz 71
Fast Track Land Reform (FTLR) 61
Framework and Guidance on Land Policy of 2009 61
Freedom Charter of 1955 56
free, prior, and consent (FPIC) process 63
FTLR *see* Fast Track Land Reform (FTLR)
Fukuyama, F. 27

Gamagara Development Forum 128
Gathii, J. 64
generalised trust 27
Ghana 38–9; mining regulation in 61
global mining industry, community engagement in 59–63
Goldthau, A. 63

governability, and social licence to operate 12
governance 42; definition of 13; theory 12
Gramson, W.A. 103
Granovetter, M. 25
Greenen, S. 141
Guatemala, gold mining companies in 30–1

Hämäläinen, R.J. 42, 94
Hanna, P. 120
Hardin, R. 27
Harvey, B. 64
Harvey, R. 3, 8
Huni, S. 62

IBMR *see* Itireleng Bakgatla Mineral Resources (IBMR)
ICMM *see* International Council on Metals and Minerals (ICMM)
ICT *see* Itumeleng Community Trust (ICT)
IDPs *see* integrated development plans (IDPs)
IIED *see* International Institute of Environment and Development (IIED)
ILO Convention 169 *see* International Labour Organization (ILO) Convention 169
Indigenisation Act, in Zimbabwe 61
Indonesia: community-mine relations 39; mining royalties in 15
influence elites 37
institutionalised trust 28
institution of traditional leadership 85–6
integrated development plans (IDPs) 4, 40–1
integrated path dependency model 92
integrative model 10–11, 16, 26
integrity-based trust 28
Inter-American Development Bank 62
intergroup contact, community-mine relations 32
Interim Protection of Informal Land Rights Act (IPILRA) 71
International Council on Metals and Minerals (ICMM) 63
International Institute of Environment and Development (IIED) 9
International Labour Organization (ILO) Convention 169 60
International Monetary Fund 62
IPILRA *see* Interim Protection of Informal Land Rights Act (IPILRA)

Itireleng Bakgatla Mineral Resources (IBMR) 71, 73
Itumeleng Community Trust (ICT) 112–13, 119–24

Jagersfontein case, social mobilisation against CDTs 112–13; community engagement 121; De Beers Sale Agreement 123; future path 123–4; Itumeleng Community Trust 119–24; Jagersfontein Developments 113, 123–4; JMMF's concerns 120, 123, 124; legitimacy and credibility 120; local concerns 119–21; local impact and benefits 122–3; social licensing process 120
Jagersfontein Developments 113, 123–4
Jagersfontein Multi-Stakeholder Mining Forum (JMMF) 120, 123, 124
Jaskoski, M. 133
Johnson, R.W. 59
John Taolo Gaetsewe Development Trust (JTG Development Trust) 112, 113, 124–9, 133
Joyce, S. 132

Kemp, D. 11, 29, 60, 95
Kenya: 2017 Mining (State Participation) Regulations in 60–1
Kgalagadi Rural Poverty Node Charitable Trust 124
Kgosi Mokgatle 86–7
Kgosi Molotlegi 89
Kgothakgothe 89, 104
Kgothatgothe 93
Kirsch, S. 6, 111, 118; *Mining Capitalism* 118
Kuruman case, social mobilisation against CDTs 113, 133; community engagement 126–7; future path 128–9; JTG Development Trust 124–9; local concerns 125–6; local impacts and benefits 127–8; *Re A Fola* Health Promotion project 128

Lahtinen, T.J. 42, 94
Langman, L. 142
large-scale mining, socio-economic effects of 31
Latin America: mining-affected communities 115–16; resource-related conflicts 115
Layton, R. 26, 42

Le Billion, P. 83
legal and regulatory frameworks 76; in Africa/African countries 60–2; in Australia 62; for community engagement 59–63; in Mexico 62; mining policy and practices 53–6, **54**; post-apartheid regimen 56–9, **58**
legitimacy 10, 28, 120, 141
Leonard, L. 4, 95
Lesetlheng community 141; sustained mobilisation in 71–5
local-level agreements (LLAs) 2; community trusts as 6–9; in mining industry 63–4; principles 64
lower-class present orientations 26
Lucas, A. 41
Luke, A. 35

Mabuza, E. 71
MACUA *see* Mining Communities United in Action (MACUA)
Maddison, S. 115
Mahoney, J. 127
Malawi, mining legislation in 7
Mandela Institute (2017) report 68
Mantashe, Gwede 70, 72, 74
Marais L. 120
Marikana massacre 32
Matebesi, S.Z. 74–5, 120
Mayan Q'eqchi' people 110
Mayes, R. 31
Mexico, mining regulatory framework in 62
Meyer, D. 103
Mineral and Energy Laws Amendment Act 47 of 1994 **58**
Mineral and Petroleum Resources Development Act (MPRDA) 28 of 2002 4, 57–9, **58**, 75, 110; controversial stipulation of 68; Section 54 of 72
Mineral and Petroleum Resources Royalty Act of 2008 58
Mineral and Petroleum Resources Royalty (Administration) Act of 2008 58
Mineral and Petroleum Royalty Act 28 of 2008 **58**
Mineral Laws Supplementary Act 10 of 1975 **54**
Mineral Policy of 1998 4
mineral regulatory regimen, in African countries 60
Mineral Resources and Petroleum Development Act of 2002 34

Mineral Rights Act 50 of 1991 55
mineral rights ownership 55
Minerals Act of 1991 **54**, 55–6
Minerals and Petroleum Resources Development Act (No. 28 of 2002) 88
Minerals Council of South Africa 69
Minerals Law of Mongolia 64
Mineras Intteruptus 11
Mining Affected Communities United in Action South Africa (MACUA-SA) 5, 129
Mining and Environmental Justice Network of South Africa (MEJCON-SA) 5
Mining Capitalism (Kirsch) 118
Mining Charter(s) 1–2, 33, 114, 147
Mining Code (2014) 61
Mining Communities United in Action (MACUA) 69–71, 141
"mining community" 7
mining–community relations *see* community–mine relations
mining companies 17; abuses by 110; in South Africa 142–3, 146
mining policy and practices, in South Africa 53–6, **54**
Mining Rights Act 20 of 1967 **54**, 55
mining royalties 5, 58, 84, 111, 133
Mining Titles Registration Act 16 of 1967 **54**, 55
Mining Titles Registration Amendment Act 24 of 2003 **58**
Minister of Indigenisation and Economic Empowerment 61
Ministry of Mineral Resources 76, 77
Minnitt, R.C.A. 56, 86
Mnwana, S. 39
Modikwa mine 2
Modipa, M.E. 97
Moffat, K. 10, 11, 26, 28, 30, 31, 33, 43, 92
Mogalakwena platinum mine 34–5
Mongolia, community-mine conflict in 64
The Moral Basis of a Backward Society (Banfield) 26
Motse, O. 120, 123
Moumo Integrated Developments (2016) 101
Mozambique, community development in 39
MPRDA of 2002 *see* Mineral and Petroleum Resources Development Act (MPRDA) 28 of 2002
Musgrave, M.K. 37

NAFTA *see* North American Free Trade Agreement (NAFTA)
Native Land Act 27 of 1913 53, **54**
Natural Oil Act 46 of 1942 **54**
natural-resource governance 53, 145; community-based 145–6; decision-making processes 146
neoliberal environmental governance 28
neo-liberalisation of customary tenure 61
The Newmont Waihi Gold Company 34
New Zealand 133; Waihi gold mining operations in 34
Ninth Alternative Indaba (2017) 5
North American Free Trade Agreement (NAFTA) 62–3

Odumosu-Ayanu, I.T. 64
Ontario, mining policies in 63
Oosthuizen, M. 142
Organisation for Economic Cooperation and Development (OECD) Guidelines on Multinational Enterprises 63
Owen, J. 11, 29, 95

Parani, R. 39
Parliamentary Portfolio Committee on Cooperative Governance and Traditional Affairs (COGTA) 85
participatory governance 3–4; in Switzerland 12
particularised trust 27–8
path dependency 14; breaking 42; EGT and 12–13; evolutionary governance theory and 42; in marketing systems 42; proponents of 13; significance of 103–4
Pavlovic, C. 94
Persha, L. 37
Peru's mining sector 110, 111, 133
Petras, J.F. 75
Petroleum Products Act 20 of 1977 **54**
Petroleum Products Act 120 of 1977 55
Phokeng, RBN case study 90; consultation and citizen participation 92–5; future pathways to social licensing 100–3, **102**; local impacts and benefits 95–100, **99, 100**; modus operandi of RBNDT/RBN 90–2
Pilanesberg Platinum Mines (PPM) 71, 73
political trust 28
post-apartheid mining regulatory regimen 56–9, **58**
PPM *see* Pilanesberg Platinum Mines (PPM)
Prakash Sethi, S. 9

156 Index

Precious Metals Act 37 of 2005 **58**
Precious Stones Act 73 of 1964 **54**, 55
Prno, J. 10
procedural fairness 11, 16, 33–6, 147; in CDT 130; of RBNDT 92, **99**
Public Trust in Business 28

racialised inequality 3
Ramaphosa, Cyril 56, 143
Ramatji, K.N. 55
Ramatlhodi, Ngoako 69
Ramphele, M. 70, 73, 74
Ravensthorpe, mining in 31
RBH *see* Royal Bafokeng Holdings (RBH)
Re A Fola Health Promotion project 128
regulatory frameworks *see* legal and regulatory frameworks
Report to Society 7–8
Reserved Minerals Development Act of 1925 **54**
Restitution of Land Rights Act 22 of 1994 54
Riofrancos, T. 115
Ripperger, T. 27
Roka Selepe 116
Rosenberg, M. 25
Royal Bafokeng Administration (RBA) 83, 93
Royal Bafokeng Development Trust (RBNDT) 13, 15, 17
Royal Bafokeng Diamonds 96–7
Royal Bafokeng Enterprise Development 89
Royal Bafokeng Holdings (RBH) 89
Royal Bafokeng Institute 89, 96
Royal Bafokeng Nation (RBN) 83–4, 104–5, 140, 143–5; BLBA and 93–4; challenges to 95; Communal Land Rights Act 88; consultation processes of 92–5; demographic context 84, *85*; governance of 89–90; history of 86–9, **87**; *Kgothakgothe* 89, 104; "locked-in" 90, 104; Minerals and Petroleum Resources Development Act 88; modus operandi of 90–2; path dependence 89–90, 103–4; quarterly performance report of 95; success of 100, 103, 105, 145; Supreme Council of 94; Tswana-speaking community of 89; unilateral decision-making 95; Vision 2035 of 101, 105; *see also* Phokeng, RBN case study
Royal Bafokeng Nation Development Trust (RBNDT) 39–40, 65, 83, 90–2,

143, 144; CEO of 91, 96, 97, 101, 102; contribution to quality of life **100**; coordinated activities of 96; procedural fairness of 92–3, **99**
Rutledge, C. 69

Section 9(2) of the Bill of Rights 57
Sheeley, R. 41
Shocker, A. 9
Sioc Community Development Trust (Sioc CDT) 124
Sishen Iron Ore Company (SIOC) 65
SLAPP *see* strategic litigation against public participation (SLAPP)
SLO *see* social licence to operate (SLO)
SLP-IDP alignment process 40
SLPs *see* social and labour plans (SLPs)
social acceptance of mining 8, 145
social and labour plans (SLPs) 1, 4, 33, 34, 40, 68
social contract 9
social infrastructure, mining impact on 30–1
social licence/licensing 10, 25, 35, 110; criticisms 29; future pathways to 100–3; and mining 14; in South Africa 142–3
social licence to operate (SLO) 1, 17, 18, 25, 43, 140, 146; community-mine relations 31–3; community trust and 29–36; conditions for 36; elite capture and 36–41; emerging paths to 131–3; governability and 12; government's role in 36; impact on social infrastructure 30–1; levels of 10; in mining industry 9–12; procedural fairness and 33–6; project-level recourse to 133–4; traditional elites and 38–40; ubiquity of 139
social mobilisation against CDTs 17, 110–11; Atok case 114–19; community trust cases in context 112–13; cross-comparitive analysis 129–31, **130, 131**; emerging paths to SLO 131–3; Jagersfontein case 119–24; Kuruman case 124–9
social responsibility agenda of mining 3
socio-economic effects 31
South African Development Act 18 of 1936 53–4, **54**
South African mining policy and practices 53–6, **54**
South African Revenue Service 58
spontaneous sociability 27

Stoker, G. 12
strategic litigation against public participation (SLAPP) 142
Subterranean Struggles (Bebbington and Bury) 125
Super Trust *see* Sioc Community Development Trust (Sioc CDT)
Sustainability Framework of the World Bank's International Financial Corporation 63
Swidler, A. 41
Switzerland, participatory governance in 12

Tapscott, C. 6
Thobatsi, J. 40
Thompson, I. 132
Thomson, I. 9, 10
Tilly, C. 27
Tracking the Trends 60
Traditional and Khoisan Leadership Bill 75–7
Traditional Courts Bill 143
traditional governance system 144
Traditional Khoisan Leadership Bill 143
Traditional Leadership and Governance Framework Act of 2003 85, 86
traditional leaders, in South Africa 86
"Transformation Trumps Sustainability" 6
Transparency International (2017) report 38
Transworld Energy Minerals (TEM) 71, 72
tribal people's land ownership rights 60
trust 25; competency-based 28; definition of 28; emotional and cognitive components of 27; evaluation 28; generalised 27; institutionalised 28; integrity-based 28; "Most-People" construct 25–6; multiple underlying assumptions of 26–9; particularised 27–8; political 28; relationships 27; three-way relationship approach 27

Trust Property Control Act (Act No. 57 of 1988) 65
2018 Draft Mining Charter 5

United Nations Declaration on the Rights of Indigenous Peoples 60
United Nations Economic Commission for Africa (UNECA) 3
United Nations Guiding Principles on Business and Human Rights 63
Uslaner, E.M. 27

Van Assche, K. 13
Vanclay, F. 120
Veltmeyer. H. 75
Verweijen, J. 141
Vivoda, V. 60
voluntary programmes 7

Waihi gold mining operations 33
ward committees 4
Watkins, S.C. 41
Wedel, J.R. 37
"We Know Our Lives Are in Danger" 141
white-owned mining companies 3
White Paper on Traditional Leadership and Governance 85–6
Williams, A. 83
Women Affected by Mining United in Action (WAMUA) 5
Wong, S. 37
World Bank 62
Wretched of the Earth (Fanon) 71

Xolobeni community 70–1, 76, 141; Amadiba Crisis Committee 72; sustained mobilisation in 71–5
Xolobeni mine 72

Zhang, A. 10, 11, 26, 28, 30, 31, 33, 43, 92
Zimbabwe 61–2; Indigenisation Act in 61
Zwane, Mosebenzi 70